Man and Environment in South-West Ireland, 4000 B.C. – A.D. 800

A Study of Man's Impact on the Development of Soil and Vegetation

Ann Lynch

BAR British Series 85
1981

B.A.R.

B.A.R., 122 Banbury Road, Oxford OX2 7BP, England

B. A. R. 85, 1981: " Man and Environment in South-West Ireland. "

ISBN 9780860541127 paperback
ISBN 9781407323589 e-book
DOI https://doi.org/10.30861/9780860541127
A catalogue record for this book is available from the British Library

This book is available at www.barpublishing.com

"But I, who have written this history, or rather fable, am doubtful about many things in this history or fable. For some of them are the figments of demons, some of them poetic imaginings, some true, some not, some for the delight of fools! "

Pronouncement made by twelfth century scribe when he had finished transcribing the saga _Táin Bó Cuailnge_ (translation from Greene, 1945, 88).

CONTENTS

Please note that additional material is available to download from www.barpublishing.com/additional-downloads.html
The original foldout has been reduced in size to match the A4 format of this book, the image is therefore not as clear as the original foldout. Please refer to the original foldout via the download for the original content.

PREFACE

The project which forms the basis of this book was carried out between 1976-1979 with the aid of the Travelling Studentship in Archaeology, awarded by the National University of Ireland. The research was undertaken under the general supervision of Prof. dr W Groenman-van Waateringe and is to be submitted as a Ph. D.dissertation to the University of Amsterdam.

Having completed preliminary work on the morphological, astronomical and metrological aspects of the stone alignments of SW Ireland in 1976 (Ní Loingsigh, 1976), I considered it worthwhile to expand this study by concentrating on the environmental and socio-economical backgrounds of these supposed Neolithic/Bronze Age monument-builders. This approach I considered to be particularly necessary because of the complete absence of modern archaeobotanical work in SW Ireland – the most recent being that of G F Mitchell in the 1950s. Initially therefore, the aims of this study were as follows:

1). By means of pollen analysis, to reconstruct the Post-Glacial vegetation history of the study area (i.e. W Cork and Kerry).

2). To examine the effect which prehistoric man had on his natural environment, with particular emphasis on the activities of the stone alignment builders.

3). To date the initiation of blanket peat growth in the area and consider the role of man as an agent responsible for the soil degeneration/podzolization which ultimately led to the onset of peat growth.

4). To recover by means of excavation, evidence of the settlements and material culture of the stone alignment builders which may in turn throw some light on the economy and social structure of the group as a whole.

5). By systemic sampling during excavation, to recover seed remains, the analysis of which may give some information about the crop agriculture and diet of the period.

6). To date the period of construction of the monuments.

In time however, modification of some of these aims was called for, with a consequent change of emphasis in the general research theme. Firstly, the pollen analysis of the monoliths of blanket peat proved to be of greater interest, but far more time-consuming, than first envisaged. One particular 3 m monolith (Cashelkeelty I) took just two days to recover in the field but its analysis took the best part of 18 months. Secondly, constraints of time and money meant that excavation had to be restricted to the period May-September 1977 and so, just two sites, Maughanasilly, Co. Cork and Cashelkeelty, Co. Kerry were fully investigated. The excavations themselves were also disappointing since the settlements associated with the monuments could not be located and the material recovered was minimal. In spite of flotation and analysis of <u>all</u> the carbonised material recovered during excavation, no seeds were identified. By the end of 1977 therefore, it was clear that the emphasis of the study would have to shift slightly from the socio-economic aspect of the alignment sites to a more general palynological study of the prehistoric period in SW Ireland, as outlined on p. 43 below.

For the purposes of this study, the term 'prehistoric period' is taken as covering the Neolithic, Bronze Age, Iron Age and ending with the Early Christian period. The boundaries of these periods are by no means clear-cut or precise and one should envisage transitionary phases while new technologies and social and economical systems were gradually replacing the old. The dating of these period thresholds has yet to be firmly established and as more radiocarbon dates become available, changes are called for in existing chronological schemes. However, in order to provide a conceptual framework within which the palaeoecological and archaeological developments may be discussed, the following chronological scheme, in round figure dates, has been adopted:

c	4000-3000 bc	earlier Neolithic
c	3000-2000 bc	later Neolithic
c	2000-1500 bc	early Bronze Age
c	1500-1200 bc	middle Bronze Age
c	1200-500 bc	late Bronze Age
c	500 bc-450 ad	Iron Age
c	450-850 ad	Early Christian

Unless otherwise stated, all radiocarbon dates bc/ad used in this study, are based on the half-life 5570 $\pm$ 30 and are uncalibrated.

All altitudes are expressed as feet and/or metres above the Ordnance Datum (OD) which is the Low Water of Spring Tides in Dublin Bay.

I would like to express my gratitude to the many people whose assistance and encouragement enabled me to see this work through to the end. My first rather tentative suggestions regarding the possibilities of environmental archaeology in SW Ireland were greatly encouraged and guided in the right direction by Prof. dr W Groenman-van Waateringe. Her continuous help and supervision, I gratefully acknowledge. Prof. M J O'Kelly helped me keep in touch with developments in Ireland and waded through the text making many useful comments and suggestions. Prof. dr T van der Hammen who acted as my co-referent, also critically read the text in its final stages. Dr Bas van Geel and Drs Jan-Peter Pals patiently coached me in the techniques of pollen and seed analysis respectively, and cheerfully braved the Irish elements in order to help with sampling in the field. In spite of his healthy scepticism regarding palynological studies, drs Albertus Voorrips spent much time producing the computer pollen diagrams and tables which were of great help and saved a lot of time during the interpretation stage of the study. Special thanks are due to John Barber who helped in every stage of the project including taking on the task of directing the Cashelkeelty excavations and supplying much useful criticism of the final text. He also bravely attempted to translate many of the

'Irishisms' into coherent English. The radiocarbon determinations were promptly supplied by Prof. dr W G Mook and G W Pearson of the Groningen and Belfast laboratories respectively. I am indebted to Bob Donker who prepared the final drawings for publication and to Trudy Domhoff and Mariejan Kooijman who were responsible for the typing. All three coped admirably with the last-minute panic and chaos. During the past three years many people prompted useful discussion regarding various aspects of the thesis, in particular Prof. G F Mitchell (Trinity College, Dublin), Prof. J J Moore (University College, Dublin), Michael Monk (University College, Cork), Dr. M O' Connell (University College Galway), Dr. J R Pilcher (Queen's University, Belfast), Prof. dr V Westhoff (Catholic University, Nijmegen) and Dr. L H van Wijngaarden-Bakker (University of Amsterdam) — to these I offer my sincere thanks. Finally, I want to mention my family and friends both in Ireland and the Netherlands who supplied constant support and encouragement even when the going was rough.

LIST OF FIGURES

LIST OF TABLES

CHAPTER 1

A SUMMARY OF THE ENVIRONMENT OF MODERN SW IRELAND

1.1 Introduction

The understanding of the ways of subsistence and economy of any group of people depends to a great extent on an understanding of their physical environment. Different environments offer different opportunities and impose varying limits on the possibilities open to people at any particular stage of culture. Similarly, the stage of cultural development can itself be measured, at least in economic terms, by the use made of the environment. In this chapter, I shall summarise the dominant aspects of the present-day environment of the study area, both physical and economical, in order to provide the background for later discussions concerning the prehistoric environment and man's adaptation to it.

In this study, the term 'environment' unless otherwise stated, may be taken as meaning physical or natural environment and includes factors such as place, climate, geology, geomorphology, soils, vegetation and time.

1.2 The Study Area

The limits of the study area are determined by the distribution of stone alignments in SW Ireland, with an alignment being defined as three or more standing stones, intervisible and in a straight line (Fig. 1). This group of alignments is one of two major concentrations in Ireland, the other being in the north, in counties Derry and Tyrone, while there are also a few isolated sites in the mid-west and east of the country (Ní Loingsigh, 1976, 23). The study area encompasses most of W Cork and Kerry, and may be delimited by a line joining Tralee/Mallow/Cork. This represents an area of approx. 2700 sq miles lying between 51°25'-52°25' North Latitude and 8°20'-10°30' West Longitude which forms the extreme southwestern tip of the island, bounded on the west and south by the Atlantic Ocean.

1.3 Geology and Geomorphology

1.3.1 Structure and Topography. During the earlier part of Lower Palaeozoic times, the area of Ireland was submerged beneath the Lower Palaeozoic Sea on whose floor sediments were accumulating. The Caledonian upheaval of the Silurian period caused the uplift of these sediments to form a landmass with a structural grain running from northeast to southwest (Mitchell, 1976, 18). During the Devonian period of the Upper Palaeozoic, the Cork and Kerry region was part of a low-lying basin in which sandstone, rich in iron oxides was formed — this is known as Old Red Sandstone. This sandstone was subsequently submerged by the Carboniferous, Upper Palaeozoic Sea and covered by a sediment of calcite-rich limestone. At the end of the Upper Palaezoic

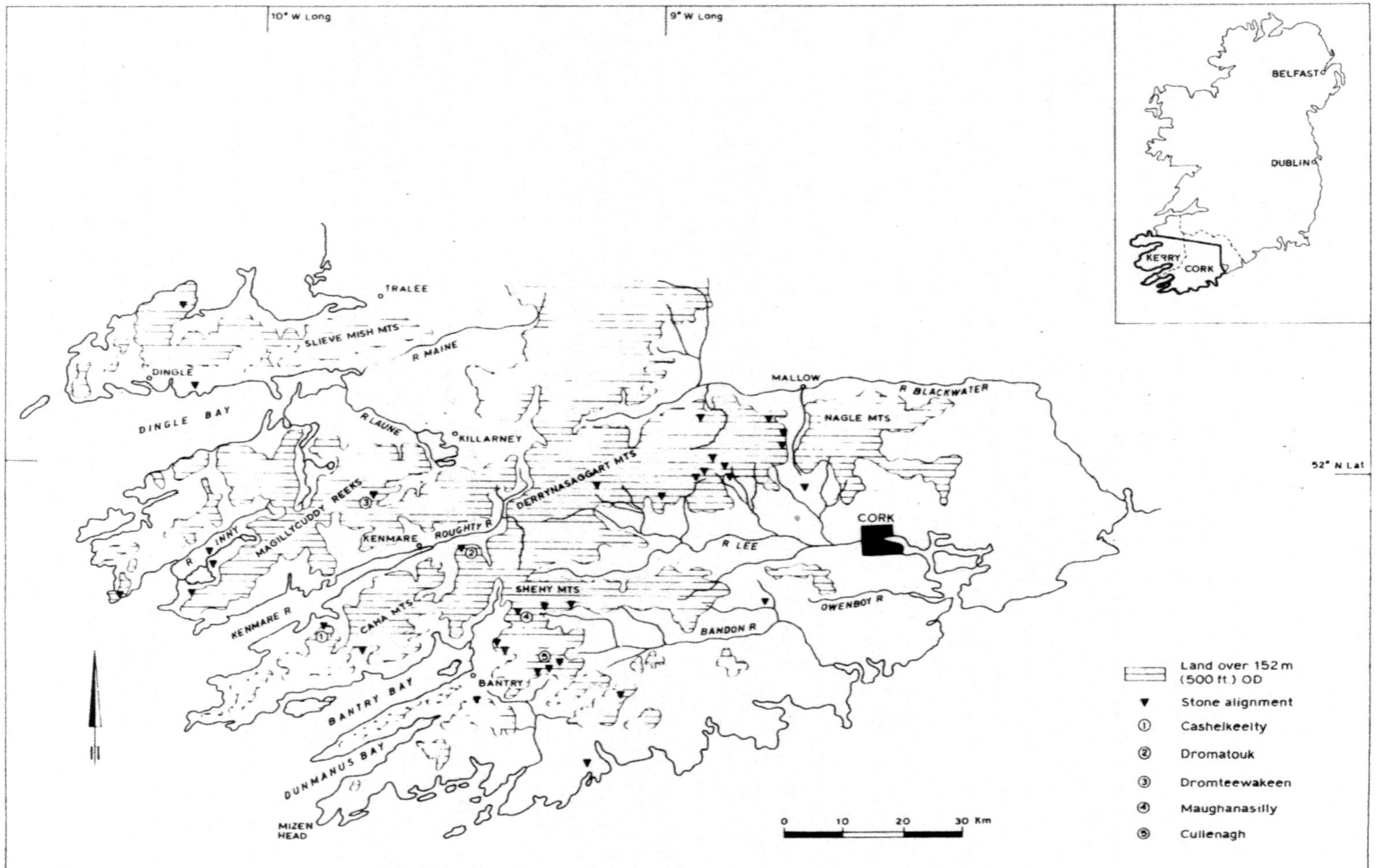

Fig 1 The study area, showing topography, drainage and distribution of stone alignments.

(about 270 million years ago) another major earth force moved from the south and raised and folded the sediments of SW Ireland in an east/west direction. This upheaval which is usually referred to as the Armorican Folding, left the Carboniferous deposits exposed to the elements and erosional processes.

The Mesozoic Era witnessed many oscillations in land and sea-levels, with alternating periods of erosion and deposition. The only deposit in SW Ireland which may be definitely assigned to this Era is a small patch of Cretaceous chalk near Killarney, Co. Kerry. Finally, the Tertiary Period of the Cenozoic Period was a period of weathering and erosion in the southwest, giving the basic land-form we have today (Fig. 1). This landform is one of parallel ridges and valleys running east/west in W Cork and Kerry. The Old Red Sandstone anticlines, from which the Carboniferous limestone has been eroded, exist as low ridges in the east and as mountains farther west, culminating in Carrauntoohil, 3414 ft (1041 m) OD, the highest mountain in Ireland. The Carboniferous limestones and shales are confined to the synclinal valleys, while in S Cork the shales have been metamorphosed to form slate which has weathered into a landscape of valleys and low ridges (Freeman, 1972). The mountainous sandstone peninsulas of Kerry and W Cork are separated by impressive rias (submerged valleys) that occupy west-pitching down-folds, largely floored by partly submerged Carboniferous rocks.

1.3.2 Glaciation. The SW of Ireland was severely affected by two major periods of glaciation during the Pleistocene, which provided the landscape with a varied microrelief. The glaciers generally scoured the uplands of rock waste, formed spectacular corries in the mountains of W Kerry, deepened existing valleys and together with its meltwaters, the ice deposited variable thicknesses of boulder clay, sand and gravel across the lowlands and mountain valleys providing a wide range of parent material for later soil development. These soils shall be described in more detail in a later section of this chapter. It is now generally accepted that many of the higher mountain peaks of Co. Kerry, plus a narrow 'corridor' stretching from the Dingle peninsula eastward to Dungarvan, probably remained ice-free during the entire Pleistocene (Mitchell, 1976, 48-63). None of the sites sampled for this study fall within this area however and so it is of little significance for the present discussion.

1.3.3 Sea-Level Changes. The details of isostatic and eustatic movements of land and sea during Pleistocene and post-Pleistocene times, are not yet fully understood. Mitchell however has built up a generalized curve for sea-level movements during the past 40,000 years (1976, 68) in which he shows that the sea reached its lowest level about 15,000 years ago (about 130 m below that of today) when the N American icemass was at its maximum. From then on, the level rose steadily until about 5000 years ago when it stood about 4 m above the present level and since then minor oscillations have occurred. It is unfortunate that these minor oscillations have not been adequately dated, since they could be of major significance in terms of land-area available for exploitation to the prehistoric settlers. In an attempt to pinpoint these oscillations, Mitchell (1976, 182) looked at the evidence provided by 'submerged forests' and 'drowned' archaeological sites, some of which occur along the Cork and Kerry coasts, and arrived at "the most tentative of conclusions .... that the sea may have been below its present level 4000 years ago and did not regain that level before 1000 years ago". If this was the case, then between c 2000 bc-c 1000 ad the coastal plain of counties Cork and Kerry would have been wider than today and so could have provided extra opportunities for settlement and exploitation. The general post-glacial rise in sea-level is largely responsible for the drowning of the lower portions of the more deeply excavated valleys, producing the spectacular indented coastline of Cork and Kerry.

1.3.4 Drainage. The trellis pattern river system of the Cork/Kerry region shows some degree of adjustment to the relief where the east/west synclines are occupied by the major east-flowing rivers such as the Blackwater, Lee, Bandon and Argideen (Fig. 1). Rather than pursue easy courses through broad synclinal lowlands to the sea however, these rivers suddenly turn south through water gaps cut through successive sandstone ridges to reach the sea in drowned estuaries. By comparison, the westward flowing rivers are insignificant in size and include the Maine and Laune which drain into the head of Dingle Bay, the Inny which flows into Ballinskelligs Bay, the Roughty into Kenmare Bay, and the Ouvane which flows into Bantry Bay.

1.4 Climate

The climate of Ireland may be summarised in three words: mild, moist and changeable. The two major factors which affect the climate of Ireland are the westerly atmospheric circulation of middle latitudes and the proximity of the North Atlantic Ocean whose temperature is markedly influenced by the warm North Atlantic Drift (Rohan, 1975).

The southwest of Ireland comprises a climatic region dominated by maritime influences and was classified by Prof. A G Tansley (1939) as 'Extreme Atlantic'. Temperatures are equable with a narrow range (about 8°C or 14°F) between summer and winter. Extreme heat and cold are virtually unknown with mean January temperatures varying from 6.5 to 7°C (44-45°F) and summer temperatures averaging 16°C (61° F). Frost is rare. Owing to the frequent cloud cover, only 27-35% of the sunshine possible throughout the year is received, a factor which frequently hinders the ripening of the harvest.

The mean annual rainfall (over 1,500 mm in the mountains, 1250-1500 mm in the lowlands) is not excessive but the number of raindays is high (200-250) with high relative humidity and feeble evaporation. Owing to depressions which generally move eastwards across Ireland, while higher pressure occurs to the south and east, moderate to strong westerly winds are common at all seasons. At Valentia Observatory on the west coast of Kerry, 62% of the wind has a westerly component and such winds frequently reach gale force (Orme, 1970, 42).

It is perhaps worth noting at this stage, the differences between the present-day climatic picture as outlined above and that of the prehistoric period (c 4000 bc-500 ad) — differences which shall be of major significance in later discussions concerning botanical and archaeological developments.

The sequence of prevailing temperatures over the past 100,000 years has been derived from various kinds of field data, e.g. biological organisms on the sea-bed, pollen analysis and oxygenisotope measurements of present-day ice caps. Lamb (1974, 207) maintains that the results obtained from these various researches are 'remarkably consistent' and the temperature changes of post-glacial times are best established for NW and Central Europe chiefly from botanical data. The conclusion arrived at is that the prevailing temperature level between 4000-1000 bc was mostly 1^{O}-2^{O}C (1.75^{O}-3.5^{O}F) above that in recent times. Analysis of anticyclone and cyclone formations have also shown that at c 2000 bc a marked belt of westerly winds was indicated in 50°-60°N which would have provided a high frequency of sunshine and quiet seas in latitudes as far north as 60°N. Alternating wet and dry periods are also evident from the occurrence of 'recurrence surfaces' dating to this period, in Irish raised bogs (Jessen, 1949, 253; Mitchell, 1956, 238-244).

After 1000 bc the prevailing temperatures seem to have dropped 'rather sharply' (Lamb 1974, 227) and for some centuries centering on 600-400 bc there was evidently a particularly inclement period with much wetness and 'frequent westerly and northerly or northwesterly storms near the Atlantic coasts'.

Gradually temperatures rose to a secondary optimum in the Early Middle Ages culminating between 1000 and 1200 ad when summer temperatures were about 1^{O}C above those of today (Rohan, 1975, 4). Finally, there was a period of short summers and long severe winters, the so-called 'Little Ice Age' between c 1430 and 1850 ad.

1.5 Soils

The present-day picture of soils in Ireland owes much to changing agricultural practices, various land drainage and reclamation projects and other cultural practices which have predominated since the fourth millennium bc when man first began his onslaught on the native vegetation. The National Soil Survey of An Foras Talúntais are at present carrying out a detailed study of Irish soils on a county basis but unfortunately, the detailed soil and land-use maps are not yet available for counties Cork and Kerry. However, from the general soil map of Ireland (An Foras Talúntais) we can identify seven dominant soil types within the study area (Fig. 2):

(a) Peaty Podzols: These intensely leached soils have developed on highly acid parent materials — mainly sandstones, slates, shales and igneous and metamorphic rocks and glacial drift containing similar debris.

(b) Brown Podzols: Orme (1970, 66) describes these leached soils as being more acid and degraded than Brown Earths although still retaining an acid mull or moder-type humus. These soils predominate in better drained areas at moderate elevations above 500 ft (c 150 m) on acid parent materials such as sandstone, shales and glacial drifts derived from such sources.

(c) Gleys: This type occurs where soils with high proportions of clay and silt are subjected to intermittent or permanent waterlogging.

(d) Blanket Peat: Blanket bogs have developed extensively on the higher slopes and hill tops where annual rainfall is greater than 1250 mm and where the evaporation rate is slow.

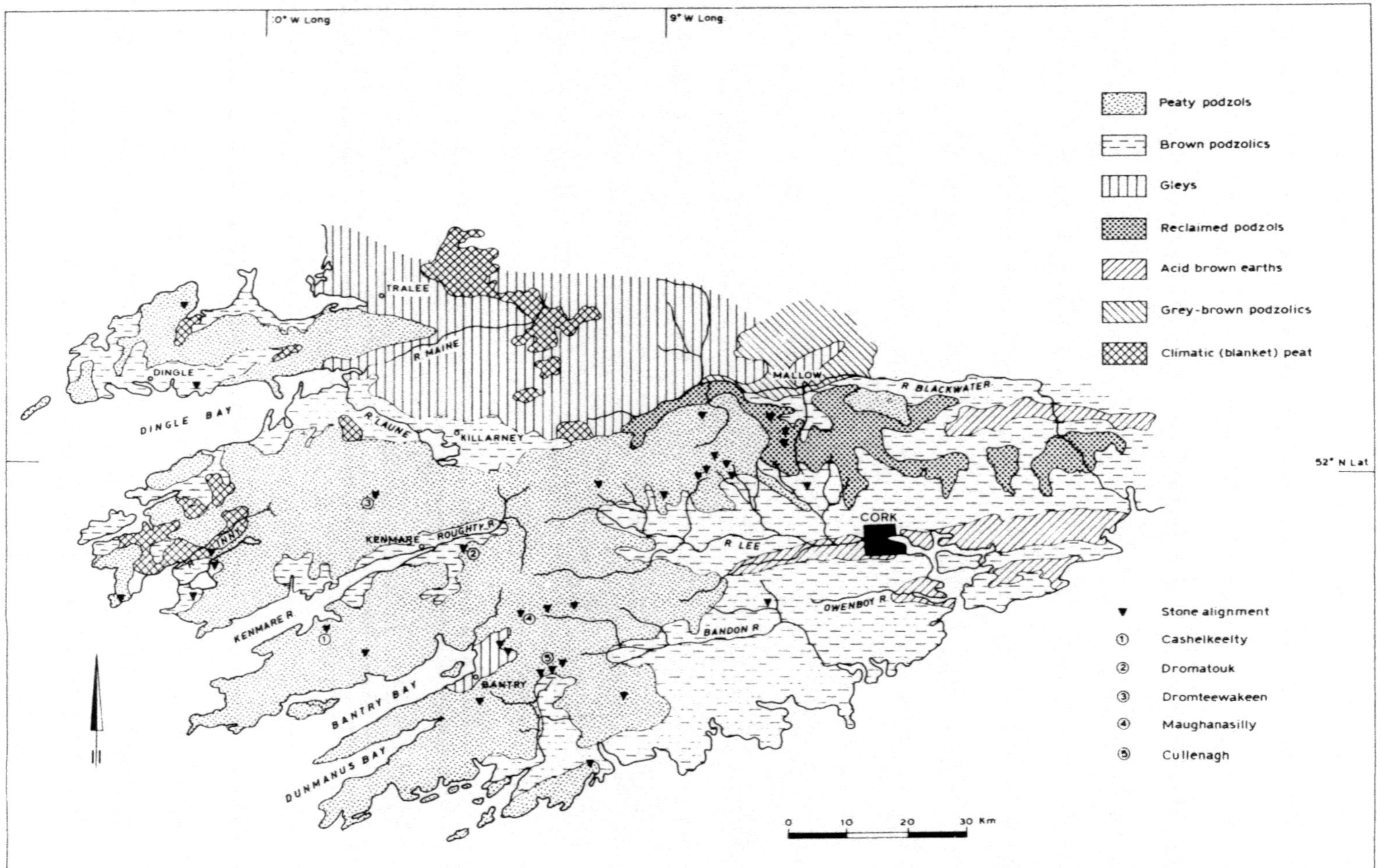

Fig 2 Soils of the study area.

(e) Reclaimed Podzols: Much of the land reclaimed in the eighteenth and early nineteenth centuries has been allowed revert to its former state (Conry 1970, 144) but quite a large tract of reclaimed peaty podzol soil is still maintained in the northern hinterland of Cork city.

(f) Acid Brown Earths: These well-drained, medium textured, relatively mature soils occur mainly on sandstone or shale parent material but may also occur on formerly base-rich drifts, that have been decalcified and base depleted through weathering and leaching. In the study area, these soils are restricted to the Lee valley and the mouth of the Owenboy river.

(g) Grey-Brown Podzols: These soils are generally well-drained, of medium-base status and of heavier texture than the Brown Earths. Some leaching takes place which results in a higher clay content in the B horizon than in the A or C. A small pocket of this soil type occurs on the northern shore of Tralee Bay, Co. Kerry.

It can be seen from Fig. 3 that about 65% of the stone alignments (including the five sampled for this study) are sited on the Peaty Podzol soil types. An understanding of the genesis of these podzolized soil profiles, now covered by blanket peat, is vital to the interpretation of the pollen diagrams presented later in this study and so, a separate chapter (chapter 5) has been devoted to this problem.

1.6 Vegetation

Following the last glaciation, with the retreat of the icesheet and consequent rise in sea-level, the Irish Sea submerged the landbridge which had connected Ireland and Britain and so formed an effective barrier to the migration of flora and fauna from dispersal centres to the east. This isolation of Ireland has resulted in a reduced flora, with about 1000 species of flowering plants and ferns native to Ireland compared with about 1500 species in Britain. There are also the interesting cases of plants which are rare or absent in Britain but which flourish in Ireland — explanations including a more favourable habitat in Ireland, have been put forward in order to explain this phenomenon (Praeger, 1977, 35; Mitchell, 1976, 95). The SW of Ireland has long been recognized as an area possessing a flora of special interest. Apart from the expected Cosmopolitan and Atlantic groups of species, the W Cork/Kerry region contains plants which are well represented in the Mediterranean basin (the Hiberno-Cantabrian element) and also plants which are widely distributed in N America (the Hiberno-American element). Species included in the Hiberno-Cantabrian element are:

Adiantum capillus-veneris (Maidenhair Fern)
Saxifraga spathularis (St. Patrick's Cabbage)
Saxifraga hirsuta (Kidney Saxifrage)
Euphorbia hyberna (Irish Spurge)
Daboecia cantabrica (St. Dabeocs Heath)
Arbutus unedo (Strawberry Tree)
Erica mackaiana (Mackey's Heath)
Erica ciliaris (Ciliate Heath)
Erica erigena (Mediterranean Heath)

Sibthorpia europaea (Cornish Moneywort)
P..guicula lusitanica (Pink Butterwort)
Pinguicula grandiflora (Greater Butterwort)
Otanthus maritimus
Simethis planifolia
Neotinea intacta (Irish Orchid)

The Hiberno-American element includes:

Hypericum canadense (St. John's Wort)
Naies flexilis (Flexible Naiad)
Sisyrinchium bermudiana (Blue-eyed Grass)
Eriocaulon aquaticum (Pipe-Wort)
Spiranthes romanzoffiana (Threefold Lady's Tresses)

(Scannell and Synnott, 1972, VI)

Five vegetational subdivisions may be recognized within the Cork/Kerry region:

1.6.1 Woodland. One of the few remnants of native oakwood to be found in Ireland, is to be seen on the mountain slopes in the vicinity of Killarney, Co. Kerry. Here the dominant tree is usually Quercus sessiliflora (sessile oak) with Betula pubescens (brown birch), Salix cinerea (fen sallow), Prunus spinosa (blackthorn) and Corylus avellana (hazel) its associates. Of the smaller trees and shrubs, Ilex aquifolium (holly) and Arbutus unedo (strawberry tree) are dominant, while the ground flora abounds in Scilla nonscripta (squill), Vaccinium myrtillus (bilberry), Luzula sylvatica (great wood-rush) and many ferns. The spread of Rhododendron ponticum in recent times is threatening to replace Ilex and Arbutus in the Killarney woodlands.

Since 1908, the Irish Forestry Commission has been implementing a policy of afforestation in areas of 'bare land', which includes portions of old demesnes not suitable for agricultural development, land above the present limit of cultivation, or heaths and dried bogs (Freeman, 1972, 73). The species favoured for this planting are Pinus sylvatica (scots pine), and the non-native Picea sitchensis (sitka spruce) and Pseudotsuga taxifolia (douglas fir). Since a large proportion of the study area may be classified as 'bare land', it has been extensively forested, with large tracts of mountainside now supporting coniferous woodland and further areas being deep-ploughed each year for future planting. By 1972, c 56 sq miles in W Cork and Kerry had been planted (Freeman, 1972, 354). The tree-line in W Cork and Kerry is extremely variable depending on protection from the south-westerly winds, but may generally be taken as lying between 600-800 ft (183-244 m) OD.

1.6.2 Peatlands. Approximately 14% of the land area of Ireland is covered by peat bogs of either the raised bog or blanket bog types. The bogs of SW Ireland (Fig. 2) belong to the blanket bog category which develops in regions where rain days exceed 250 a year and rainfall is greater than 1250 mm per annum. They may occur on level land or on slopes up to 25°. Where land is undulating the peat surface tends to conform to the contours of the underlying soil, but by filling in hollows the overall effect is a smooth landscape with gentle curves. The peat is usually two to three metres deep.

Blanket bog occurs in two forms, a mountain type which is found at altitudes greater than 1200 ft (400 m) and a western lowland form. The mountain blanket bog supports a vegetation of Eriophorum angustifolium (cotton grass), Scirpus caespitosus (deer grass), Narthecium ossifragum (bog asphodel) and widespread Myrica gale (bog myrtle). The vegetation of the western lowland bogs is rather similar but supports a greater abundance of Molinia caerulea (purple moor grass) and Schoenus nigricans (black bog rush).

1.6.3 Coasts. The coastline vegetation of the study area is strongly influenced by exposure to the prevailing winds and many of the species which flourish on the east coast do not appear in the west (Praeger, 1977, 61). In the more sheltered areas sand-dunes are backed by stable mossgrown sandy ground while in other more exposed parts there are stretches of coastline quite bare of vegetation. According to Praeger (1977, 62) 'semi-exposure to the Atlantic sometimes produces the phenomenon of a salt-marsh flora at the foot of a storm-beach of boulders.' The flora of the sea-rocks and cliffs includes Lavatera (mallow) and Spergularia rupicola (rock spurrey). Finally, there is the unusual phenomenon of the Plantago Sward which occurs on the wind-swept clifftops and exposed slopes. This has been described by Praeger (1977, 63) as 'a very close dwarf maritime sward .... composed mainly, often almost wholly, of Plantago maritima and Plantago coronopus, growing extremely small.'

1.6.4 Mountains. The Irish Alpine flora in general is poor. The 'Highland' plants of the SW are depressed in altitude because of the conditions prevailing on the Atlantic sea-board, for example, in Kerry, plants such as Sedum roseum (roseroot), Galium boreale (northern bedstraw) and Juniperus sibirica (juniper) all have their lower limit between sea-level and 70 ft (21 m). On the other hand, the vegetation of even the highest mountains is made up mainly of common lowland plants.

1.6.5 Lakes. Apart from the famous Lakes of Killarney, the study area abounds in small lakes and mountain tarns up to about 2500 ft (762 m). The pH factor is of major importance in determining the flora of these lakes, and apart from the Lower Lake at Killarney, which lies on limestone, the majority of the smaller lakes have acid waters. The vegetation of these lakes and tarns includes Elatine lexandra (waterwort), Callitriche intermedia (starwort), Lobelia dortmanna (water lobelia), Juncus bulbosus (bulbous rush), Isoetes lacustris (quillwort) and Nitella translucens.

1.7 Rural Economy

A consideration of the present-day subsistence economy of W Cork and Kerry is of major significance when trying to reconstruct the response of prehistoric man to his environment, in terms of settlement pattern and agricultural activities. Factors such as relief, drainage and to a certain extent climate, have remained unchanged since prehistoric times and so provide a certain common denominator with the modern situation. Such a comparison is even more valid when we consider that the study area has escaped industrialization of any consequence and has remained relatively undeveloped over the centuries.

As with the remainder of Ireland, W Cork and Kerry experienced a rapid growth in population in the eighteenth and early nineteenth centuries culminating in a figure of c 608,000 in 1841 (Kerry 294,000 and W Cork c 314,000).

The potato Famine of the 1840s brought about a drastic reduction in population through death and immigration and this drain on population continued right up until the 1970s when the first signs of stabilisation became evident. In 1971, Co. Kerry had 113, 000 inhabitants and W Cork had c. 201, 000 (excluding Cork city), representing a decline of 37% and c 36% respectively (Freeman, 1972, 336, 358). Town and village populations account for approximately one-third of the whole in Kerry and one-fifth of that in W Cork which means that about 70% of the total population lives in dispersed habitations. The settlement pattern shows a fairly even distribution of population in most of the lowlands with some 'ribbons' of habitation and cultivation on hillsides or in valleys and scattered groups of houses in the middle of the bogs. The population density per square mile of improved land varies between 100 and 200.

The study area may be divided into four sub-regions of differing subsistence patterns:

1.7.1 The Lowlands of North and Central Kerry. These lowlands are on carboniferous limestone with much peat development along the alluvial river plains. About 50% of farms in this area are of 30-100 acres (12-41 hectares), 10% of more than 100 acres (41 hectares) and the rest of less than 30 acres (12 hectares). The economic staples are store and dairy cattle, pigs and poultry, with crops grown mainly for fodder (Freeman, 1972, 360). Grassland in pasture and meadows predominates and a typical crop rotation would be as follows: oats on lea ground, followed in the second year by potatoes and roots, then by oats with which grass seed is planted and in the fourth year hay after which the field reverts once more to pasture for several years. The increasing importance of milch cows is reflected in the numerous creameries which have sprung up in the area in the past decade.

The density of the rural population averages 100-200 per square mile of improved land and over 80 per square mile of the entire area. The major towns of Listowel, Tralee and Killarney are predominantly market towns with tourism becoming increasingly important in the latter.

1.7.2 The Uplands of North and Central Kerry. These uplands of sandstone, shales and grits have relatively level surfaces at 600-800 ft, 1000 ft and 1200 ft and the river valleys of the Laune, Maine and Feale dissect these plateaux. The density of population is about 50-60 to the square mile with most of the settlement in the more open valleys and the remainder restricted to clearing on the hillsides or on the extensive tracts of bogland in the area. Most of the farmers have small holdings with arable farming being restricted to the growing of potatoes and oats. Rotation takes the form of two year's potatoes, one years' oats and then a return to grassland. Cash staples are bacon, butter and eggs with the better farms concentrating on store and dairy farming. Many of the upland farms which were occupied in the past century are now abandoned and the land has been allowed revert to rough grazing.

1.7.3 The Peninsulas. The six peninsulas to the southwest present a single, rather depressing, picture of subsistence. From north to south, these peninsulas are the Kerry Head Peninsula, the Dingle Peninsula, the Ivereagh Peninsula, the Caha Peninsula, Sheep's Head Peninsula and Mizen Head Peninsula (Fig. 1).

The density of population averages about 50 per square mile but on the improved land it is mainly around 150-200. The settlements are dispersed along the coastal plains with some scattered houses to be found along the narrow river valleys in the mountainous interiors. The average farming unit is 30 acres and extensive areas have been allowed revert to rough pasture, especially on the boglands and towards the upper limits of cultivation. Any form of rotation practised is limited to the growth of potatoes for two or three years followed by oats or green crops. Some fields are held in meadow or pasture with both cattle and sheep pastured on the rough mountain and moor grazing lands which are mainly held in common. On the smaller, more mountainous farms, the spade is often the only implement used and seaweed and sand are still used as manures. Cash products include butter, eggs, poultry, pigs, calves, sheep and wool — in general the products of mixed farming.

Fishing was a thriving industry along the whole southwestern coast until the past few decades when the effects of over-fishing began to make themselves evident. Today trawler fishing is mostly restricted to Dingle and Tralee and there is still much oyster, mussel and lobster fishing, geared to the Continental market.

1.7.4 The Skibbereen Lowlands. These southwestern lowlands are drained by the river Ilen and have a density of population of 60 per square mile, with 84 on the improved land (Freeman, 1972, 370). The average agricultural holding is 40-50 acres and crops including oats, mangolds, sugar-beet and wheat cover some 17% of the land. The major emphasis is however, on producing store cattle and milk for the creameries. This area was one of those worst hit by the decline of the fishing industry.

1.7.5 The Cork Ridge and Valley System. This final sub-division of the study area encompasses the area extending eastward from that already described to the artificial study-boundary running from Cork city northward to Mallow, an area which is in fact part of a larger structural unit which extends through E Cork into Co. Waterford. This is an area of well-defined Old Red Sandstone ridges separated by the valleys of the Blackwater, Lee, Bandon, Owenboy and Argideen rivers.

In contrast to the other sub-divisions, the pattern of settlement here is relatively even, with the few gaps corresponding with the upland areas. Freeman (1972, 336) attributes this pattern to the large percentage of medium size farms (75-85 acres, 31-35 hectares). Density of population averages 100 per square mile with the frontier of agricultural settlement lying between 600-1000 ft.

Due to the farmers' reliance on a return of some sort from cattle, the land use pattern is pasture first, meadow next and arable crops last. Mixed farming is therefore characteristic everywhere with stock production as the main aim. The area to the north of Cork city is particularly geared to supplying the city market with vegetables and dairy products. Sugar-beet is grown for the sugar factory at Mallow. More than 50% of this area is improved land and the limit of cultivation is quite high with less land being allowed revert to rough pasture than in the other sub-divisions.

1.8 Summary

The study area of Cork and Kerry forms a well-defined unit in terms of structure, climate, soils and vegetation. It is essentially an area of east/west mountain ranges separated by broad valleys, which in the extreme west have been submerged to form deep inlets or rias. The major rivers flow west-east and then turn south to enter the sea in broad estuaries. The climate of this part of Ireland is extremely mild due to the influence of the North Atlantic Drift, and the number of raindays per annum is high.

The soils are composed mostly of glacial boulder clay derived predominantly from sandstone and limestone parent materials. Since the last glaciation, these soils have developed in various ways, with the process of podzolization being most widespread in this area. Large tracts of these podzolized soils are now covered by blanket bog. The vegetation of W Cork and Kerry is unique in many ways. The mild moist climate supports a sub-tropical vegetation in places and many N American and Mediterranean species also thrive here.

Finally, the present-day subsistence pattern in the study area is basically one of mixed agriculture. Farms are privately owned but small (averaging 30-50 acres or 12-20 hectares). Much of the marginal land which had been reclaimed in the nineteenth century has been allowed revert to rough grazing. Dairy and store cattle, pigs and poultry are the cash items while the staple crops are oats and potatoes. The towns of the region serve as market centres and tourism is of growing importance.

CHAPTER 2

A SUMMARY OF THE PREHISTORIC ARCHAEOLOGY OF SW IRELAND

2.1 Introduction

The framework of prehistoric studies is based very definitely on purely archaeological evidence. The material remains of man's past have for long occupied archaeologists with the problems of creating chronological schemes and typologies, and with reconstructing the physical attributes of his everyday life, in terms of houses, tombs, pottery, tools, weapons and ornaments. Recent years have seen a dramatic change in approach, and the emphasis is now on the social, economic and even psychological aspects of early societies with the attendant variety of techniques, theories and hypotheses. Such an approach was long overdue: no longer are archaeological artefacts (i.e. any material remnant of man's past) considered in isolation but viewed rather as integral parts of a larger cultural system. Before embarking on such ambitious projects however, it is vital that the reliability and limitations of the archaeological data used for much of the hypothesis testing, is fully recognized. Only then can a proper evaluation be made of the significance of results and conclusions which may arise from such a piece of research. Since this study is an attempt at reconstructing the physical environment of early man in SW Ireland and examining the man-land relationship of the prehistoric period, this chapter will be devoted to outlining the archaeological picture of the period in question, thereby facilitating an integration of both aspects in later discussion. This outline shall be restricted to the study area with references to other parts of Ireland where relevant.

Archaeological research in SW Ireland has been very much in the "traditional" mode. Excavations have been carried out in isolation with little attention having been paid to research design or the tackling of salient problems and questions. In recent years, rescue excavation has dominated the scene which has resulted in a further diversification of site types examined and standard of data retrieved. Surveys of specific monument types, together with general regional and local surveys are being undertaken, and in time, should yield much useful information for further work on topics such as patterns and forms of settlement.

Before considering the archaeological remains, it is necessary to consider the problems of differential destruction of evidence, both on a macro-scale where whole sites are concerned and on a micro-scale when we consider small finds recovered during excavation. The possibility of being left with a biased sample as a result of selective destruction is a very real one and as Bradley (1978, 3) points out, "the continuous pressure placed on the best resources means that the quality of the remaining material can be in inverse proportion to its original significance." Fortunately, the problem of destruction in the study area has not been too great. The saving factors have been the very mountainous and therefore relatively inaccessible nature of the landscape and the marginal agriculture practised which does not call for extensive land clearance and reclamation. In recent years, the forestry plantations on the uplands and building development in the eastern lowland area have been the major threats to archaeological sites. However, an increasing awareness amongst the general public of the importance of archaeological monuments, has meant that many have been fully excavated before destruction. Perhaps of greater importance than physical destruction, is the masking effect of the very extensive tracts of blanket bog which cover much of the study area. This peat cover which in places can be several metres thick, results in a very real bias towards remains which survive to a height of c 1 m or more, and this in effect means that the more prominent burial monuments of a period may be evident, but all traces of the more ephemeral houses and other domestic structures are totally concealed. Large scale projects of aerial photography combined with phosphate analysis of the underlying soils are needed to rectify this situation, but until then, the distribution of the more obvious remains shall have to be taken as representing, within limits, the original distribution of settlement.

Even if fully excavated, a site will very likely produce a sample of finds unrepresentative of the original assemblage of the community being studied. The generally acid soil conditions (particularly in the peatlands) of W Cork and Kerry militate against the survival of bone which creates an enormous gap in any subsequent study of the economy or physical anthropological aspects of a society. The remarkable dearth of pottery in this area has also been blamed on the soil conditions but further excavation and experiment is needed before this claim can be substantiated. In spite of the limitations just outlined, a study such as this, which attempts to relate the archaeological evidence with the environmental, should not be without its uses. Many gaps in our knowledge concerning the prehistoric period, may be filled by palaeoecological investigations and the latter become even more important when archaeological data are not readily available. Some attempt must be made to answer questions posed and in time these answers may combine to give a well-balanced view of the prehistory of W Cork and Kerry. The following outline shall be dealt with under the broad headings of Neolithic (c 4000-2000 bc), Bronze Age (c 2000-500 bc), Iron Age (c 500bc-450 ad), and the Early Christian Period (c 450-c 1100 ad). The date ranges should be regarded as nothing more than guidelines and the complexities of recognizing and dating the transition between chronological periods will be discussed in a later chapter.

2.2 Neolithic (c 4000-2000 bc)

Evidence for Neolithic settlement in SW Ireland is tantalisingly elusive. While the northern and eastern parts of the country exhibit signs of widespread Neolithic activity in the form of court-tombs, passage-tombs and portal-tombs, such megalithic remains are absent from the southwest with the exception of two possible portal-tombs on the coast of Cork (Ó Nualláin, 1979, 10). The problems of dating, origins and affinities of these three megalith types have for decades occupied Irish archaeologists and the most recent summaries of the various views may be read in Ó Nualláin (1979), de Valera (1979) and

O' Kelly (forthcoming). Basically, the de Valera and Ó Nualláin stand is that the three tomb types resulted from two separate invasions by colonists from W Europe with differing burial customs. The first colonists to arrive (c 3000 bc) were the court-tomb builders who landed on the northwest coast, from whence they spread eastward, finally reaching SW Scotland and the Isle of Man. Not long afterwards the passage-tomb builders landed on the east coast and "spread in strength across the northern edge of the central plain" (de Valera, 1979, 6). Finally, the portal-tomb is regarded as having evolved in Ireland from court-tomb. Such theories regarding the Neolithic settlement in Ireland are based on two assumptions: (i) that there was no indigenous Mesolithic population in Ireland and therefore that Neolithic colonists were faced with a "virtually if not totally virgin country" (de Valera, 1979, 1); (ii) the dating of the monuments indicate a court-tomb, passage-tomb, portal-tomb sequence, with overlaps. O' Kelly (forthcoming) however undermines these assumptions by considering the now extensive evidence for a strong Mesolithic presence in N Ireland (Woodman, 1974, 1976, 1978) which in his opinion could have provided the necessary stimulus and foreign contacts for a gradual and complex transition to a Neolithic way of life without the need for any 'invasion' of colonists. Secondly, the more recent radiocarbon dates, particularly those centering on 3100 bc from the passage-tomb cemetery at Carrowmore, Co. Sligo, are pointing more towards a contemporaneity in construction of the three tomb types which could be the result of completely native influences. Far more excavations and dating programmes shall have to be undertaken before the contribution of foreigner and native to the concept of such tombs, can be recognized with any certainty. Whatever their origins, these Neolithic settlers do not appear to have penetrated as far as the southwestern corner of the island, an area which seems to have remained devoid of any settlement until the late Neolithic/early Bronze Age period. This brings us to the problem of the wedge-tombs.

There are 387 of the relatively simple wedge-shaped megalithic tombs in Ireland. Their distribution shows a marked bias to the entire western half of the island with the major concentrations occurring in Co. Clare and the W Cork and Kerry region (Fig. 3). The 70 or so wedge-tombs in the study area are concentrated around the headwaters of the Lee and Bandon rivers and around the coastal strips of the peninsulas, a distribution which coincides very closely with that of the stone alignment/stone circle complex (see below p. 18). The major problem regarding the wedge-tombs is, do they belong to a native Irish late Neolithic development of the megalithic tradition as O' Kelly (forthcoming) suggests or do they represent the arrival of yet another group of de Valera' s and Ó Nualláin' s colonists, this time coming from the Breton area of France during the early Bronze Age period (de Valera, 1979, 11; Ó Nualláin, 1979, 15)? The evidence, as it stands, is tenuous. Firstly, out of a total of 387 tombs, only 19 have been excavated, 8 tombs produced no pottery, 8 produced Beaker pottery, tanged and barbed arrow-heads were found in 3 and "metal finds from four sites may be primary deposits" (Ó Nualláin, 1979, 15). Any attempt to come to plausible conclusions from such a tiny statistical sample is dangerous, especially when both Neolithic and Bronze Age elements are equally unreliable. The Cork and Kerry excavated tombs have been particularly poor in finds, for example

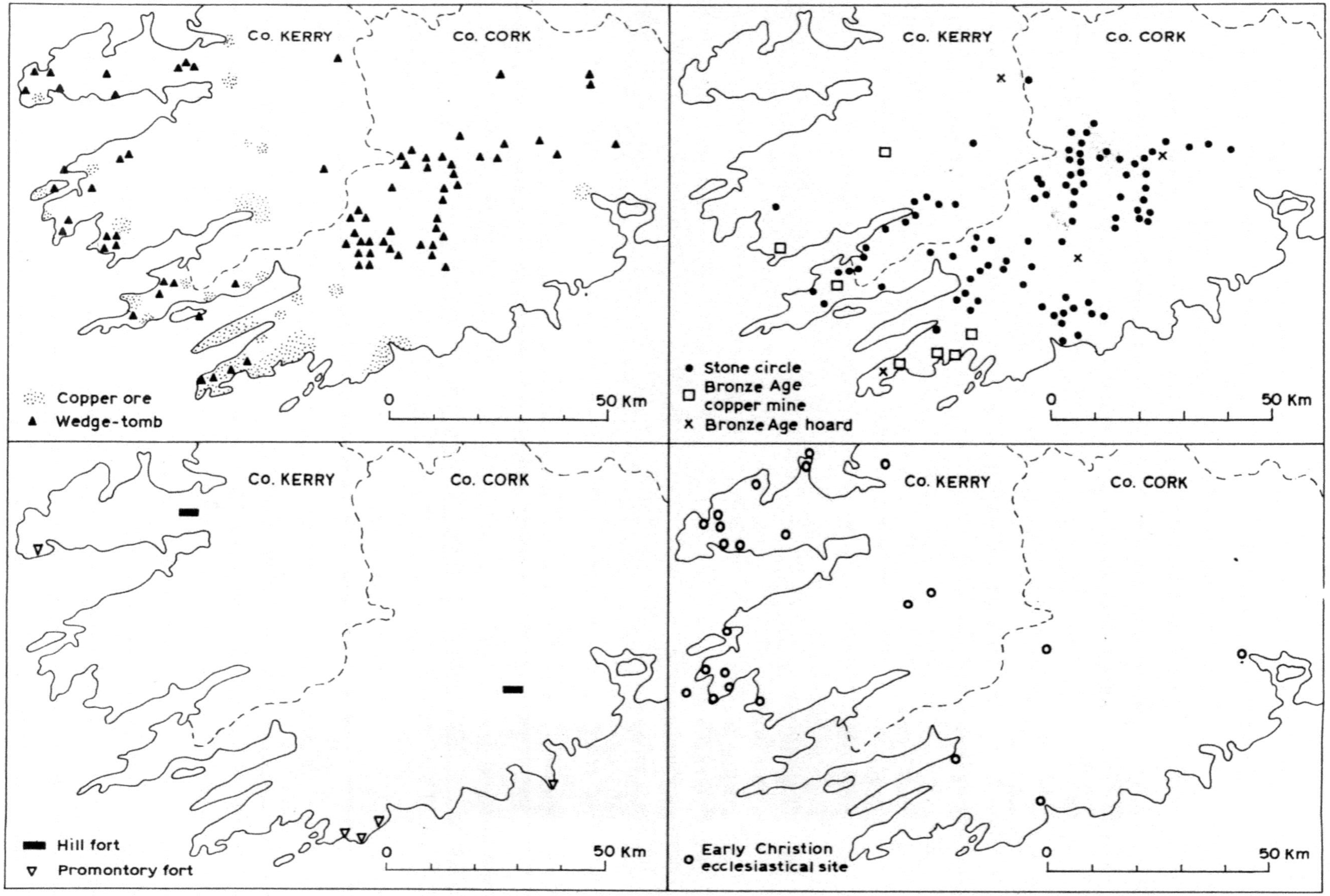

Fig. 3 Distribution of major prehistoric sites in the study area - after Ó Nualláin 1975, Jackson 1979, Hughes and Hamlin 1977.

the four excavated most recently by Herity (1970) in the Ballinskelligs/ Waterville area of Co. Kerry produced "no grave furniture".

The coincidence of copper ore deposits in the W Cork and Kerry region with the wedge-tomb distribution is often cited as supportive evidence for the Bronze Age dating of the tombs, especially since these deposits were extensively worked in ancient times and a radiocarbon date of 1500 ± 120 bc (VR1-66) has been obtained from one of the mines at Mount Gabriel, Co. Cork. Such a wedge-tomb/copper mining association may well be determined in the future, but right now there is no evidence to support it. It should also be remembered that where the densest concentration of wedge-tombs occurs, i.e. the northern half of Co. Clare, there are no ore deposits whatsoever.

A final aspect of these wedge-tomb builders is their obvious avoidance of heavy lowland soils which has been taken by de Valera (1979, 123) and Ó Nualláin (1979, 15) as indicating a subsistence pattern based on stock raising rather than crop agriculture. The concentration of tombs in N Clare is that most often cited in support of this view, where according to de Valera (1979, 123) the surrounding limestone-based soils which "are capable of feeding cattle throughout the year would have been especially valuable to primitive farmers who would not have had hay or root crops for winter fodder". The fact that this limestone area would have provided good winter grazing could certainly have attracted the high numbers of settlers to the area but to claim that this pastoral advantage precludes the possible existence of crop agriculture is rather premature since little or no archaeobotanical work has been carried out in the vicinity of the tombs. The remaining 300 tombs, while retaining an upland distribution, are scattered over soils of varying agricultural potential where local topographical advantages are often exploited. The key to solving problems regarding the subsistence of the wedge-tomb builders lies in their settlements which so far have succeeded in eluding the archaeologists. The many stretches of ancient field-fence systems often visible in the vicinity of wedge-tombs in Co. Kerry could very well belong to the same culture and provide further information regarding extent and usage of land resources. Similar peat-covered field-fences are at present being excavated in Co. Mayo and Co. Sligo (Caulfield, 1978) for which Neolithic and Bronze Age dates are being postulated.

On the basis of present archaeological evidence therefore, the first settlers in the study area were the wedge-tomb builders who arrived sometime in late Neolithic or early Bronze Age times (i.e. centering on 2000 bc). Where these colonists originated is uncertain but the clear distribution of sites around the headwaters of the rivers Lee and Bandon and along the peninsula coasts suggests that the area was penetrated from the south and west and that the new settlers arrived by sea. The subsistence economy of these peoples is as yet unknown, they may have been farmers seeking new lands to work or they may have had more interest in the potential offered by the copper ore deposits of the area. Their settlements have not been traced and the nearest identified Neolithic/Bronze Age settlement is approximately 100 miles (c 160 km) to the northeast at Lough Gur, Co. Limerick.

2.3 Bronze Age (c 2000-500 bc)

2.3.1 Burial/Ritual Monuments. In contrast with Neolithic times, the Bronze Age in W Cork and Kerry was a period of intense activity. Indications of this activity have registered strongly on the pollen analytical record (see below p. 132) and the archaeological remains include a complex of stone circles, stone alignments, standing stones, boulder dolmens and cairns. Any one of these five burial/ritual monuments types may be found in close proximity to one or more of the remainder and what excavation has been carried out suggests that all five are broadly contemporary. The stone circles have been studied in detail by Barber (1972, 1973), the stone alignments by Lynch (Ni Loingsigh, 1976) while the complex as a whole has been dealt with by Ó Nualláin (1975).

There are two major concentrations of stone alignments in Ireland—one in the Derry/Tyrone region of Northern Ireland and the other is the W Cork/Kerry group which consists of approximately 68 monuments. Three morphological differences may be noted between the northern and southern alignment groups: (i) the average height of stones in the northern alignments is 1-1.5 m lower than that for the southern group; (ii) the northern alignments are found almost exclusively in close proximity to stone circles and cairns while the majority of the southern sites are found in isolation; and (iii) the maximum number of stones in a W Cork/Kerry alignment is 6 while a northern site can have as many as 24.

Only two sites of the southern group have been excavated, Maughanasilly, Co. Cork and Cashelkeelty, Co. Kerry (see chapter 6 below) and while finds were minimal, the radiocarbon dates and pollen analyses suggest a late Bronze Age date. The Beaghmore, Co. Tyrone, complex of circles and alignments has been given the date bracket of 1535 ± 55 to 775 ± 55 bc for the construction of its monuments (Pilcher, 1969, 89). Two other such complexes of the northern group have been excavated, Castledamph, Co. Tyrone (Davies, 1938) and Drumskinny, Co. Fermanagh (Waterman, 1964), but no reliable dating evidence was recovered. The problem of origins and affinities of the W Cork and Kerry alignments is therefore difficult. Apart from relating the northern Irish alignments to the English and Welsh sites (on the basis of morphology) and comparing the southern Irish group with the Scottish alignments, there are no grounds for placing one group earlier or later than the other.

The distribution of stone alignments within the study area (Fig. 1) corresponds remarkably well with that of the wedge-tombs, as has already been noted (p. 15) and also with that of the stone circles. The only alignment outside the general distribution is that at Beal Middle in northern Co. Kerry. The preferred altitudes range between 70-830 ft (64-799 m) OD which means that there is a marked concentration of sites on the hillsides and the sides of valleys with just a small number of isolated sites on the valley floors. As with the wedges, there is clustering of sites around headwaters of rivers, particularly the Lee and its tributaries and around the coastal strips of the peninsulas. This distribution means that over 90% of the alignments are sited on soils which are now peaty podzols but in Bronze Age times would have been light stony lithosols and hence easily worked by primitive agriculturalists.

The stone circles of W Cork and Kerry which number c 90 are also one of two distinct Irish groups, the other being in Counties Tyrone, Derry and Fermanagh in N Ireland. The W Cork/Kerry type of stone circle has been described by Ó Nualláin (1975, 84) as "a ring of free-standing orthostats, uneven in number and symmetrically arranged so that one stone, the axial stone is set directly opposite to stones marking the entrance to the circle". This 'axial' stone of Ó Nualláin's is more frequently referred to as the 'recumbent' stone of a circle since it is normally placed on its side rather than on end and has a straight horizontal upper edge. Almost half of the circles have five stones and an internal diameter of c 2.50 m while the remainder (referred to as multiple-stone circles) can have up to 15 uprights and internal diameters of maximum 17 m.

As can be seen from Fig. 3 these circles are grouped along the highlands which flank the western Lee basin, in the Skibbereen lowland area and along the coastal strip of the Beara peninsula. The one point of contrast with the alignment distribution is the obvious avoidance of the Dingle and Ivereagh peninsulas. A total of 6 of these stone circles have been excavated and these are Kealkil (Ó Riordáin, 1939), Drombeg, Bohonagh and Reenascreena South (Fahy, 1959; 1961; 1962) all in Co. Cork and the two circles at Cashelkeelty, Co. Kerry, the excavation of which is described in chapter 6 of this study. The paucity of datable finds from these sites is unfortunate. Kealkil and Reenascreena South produced no artefacts at all, while some nondescript flints were recovered at Drombeg and Bohonagh. Drombeg also produced parts of a coarse flat-bottomed pot which Fahy the excavator likened to the Lough Gur Class II wares and hence "tentatively" assigned it to the early or middle Bronze Age. However, charcoal from a central burial deposit has given a radiocarbon date of 600 ± 120 ad (D-62) which may mean that the burial is a secondary intrusion in the circle since its date compares favourably with that obtained for a nearby cooking place or <u>fulacht fiadh</u> (368-608 ad). Fahy also put forward the alternative explanation that the coarse pottery type continued into late Bronze Age and Iron Age times and that the whole Drombeg monument may belong to that late period. As shall be seen later, the excavation of the Cashelkeelty circles has produced radiocarbon dates centering on 845 bc and the evidence of finds and the pollen analysis also point to a late Bronze Age date.

The monument type most often found close to and often within stone circles is the boulder dolmen (Fig. 3). The boulder dolmens are very simple monuments consisting of a single capstone (c 2 m long) resting on very low supporting stones. The term 'boulder burial' has recently been adopted by Ó Nualláin when referring to these monuments, on the basis of the excavation of the Bohonagh site (Fahy, 1961) which produced a cremated burial. However, since this is the only site excavated out of a total of c 50 it may not be representative of the group as a whole.

Other monuments occurring within this general complex are small cairns (e.g. Kealkil, Co. Cork and Lissyviggeen, Co. Kerry) and single or pairs of standing stones. The only cairn excavated has been that at Kealkil (Ó Riordáin, 1939) but it produced no finds or burials.

In spite of the limited excavation which has been carried out, it seems valid for the moment to regard the stone circle/stone alignment complex as part of a single cultural unit. The orientation of both the circles and alignments has been examined in detail by Barber (1973) and Ní Loingsigh (1976) respectively. The results indicated a consistent northeasterly/southwesterly orientation with the more obvious events of the solar and lunar cycles being possible targets of orientation. A similar tendency for the orientation of the entrances of wedge-tombs has been noted by de Valera and Ó Nualláin (1964, 113; 1972, 158).

The precise use or purpose of the monuments remains obscure. Cremation burials have been found in four multiple-stone circles, Reenascreena South, Bohonagh, Drombeg and Cashelkeelty, while an inhumation burial may have been associated with the Cashelkeelty alignment (see below p. 66). The Bohonagh boulder dolmen also covered a cremation burial. Fahy however regarded the circles as ritual rather than sepulchral monuments with the burials being simply dedicatory and not providing a <u>raison d'être</u> for the circles (Fahy, 1962, 66). His argument was that if the circles were meant as sepulchral monuments then one would expect them to have been re-used after erection but there is no evidence that such was the case.

Finally, there is the question of why the study area, which had been more or less completely ignored during Neolithic times is suddenly the focus of attention during the Bronze Age. The fact that prehistoric copper mines, some apparently of Bronze Age date, occur along the southern coastal belt of Co. Cork and in the Killarney district of Co. Kerry (Fig. 3), plus the coincidence of the distribution of many of the Bronze Age monuments with copper ore deposits, has led to the oft-expressed theory that the wedge-tomb builders and the stone circle/alignment people were attracted to the area by its potential for copper exploitation. This is certainly an attractive theory, especially when one sees the coincidence of monuments with ore deposits on the Iveragh, Beara and Mizen Head peninsulas. There are two problems which should be kept in mind however; (i) the absence of copper or bronze artefacts from the excavated sites and (ii) the dense concentrations of wedges, circles and alignments along the Lee and Bandon basins where there are no ore deposits. Discussion of this whole problem of explanation of settlement/burial pattern and the subsistence economy of the Bronze Age peoples in the study area is best left until later (chapter 8) when the environmental evidence has also been considered.

Before leaving the question of Bronze Age burial traditions it is necessary to mention the cist graves and pit or urn burials of the early/middle Bronze Age. The cists have been dealt with by Waddell (1970) and the urns by Kavanagh (1973). Over 600 cists which may be dated to the Bronze Age are known from Ireland. The majority (c 88%) of these are short cists (c 0.80 m long x 0.50 m wide) usually containing a single cremation burial although inhumations may also occur. Waddell (1970, 97) notes that c 52% of all known cists contained one or more pottery vessels. Foodvessel pottery types were by far the most common occurring in 217 cists, 47 contained cinerary urns, five contained pygmy cups and 59 contained "unidentified urns". Approximately 46% of Irish cists occur in cemeteries which may either be 'flat', with no surface indicators marking the positions of the cists or in 'cemetery mounds' where earthen mounds "may cover, incorporate, or have inserted in them at least three separate graves". The distribution of cists covers every county in Ireland with the exception of Kerry and only 4 occur in the study area, in the Blackwater and Lee valleys (see distribution map in O'Kelly and Shee, 1974, 83). A similar eastern bias is seen in the distribution of simple urn burials which consist basically of a cinerary urn inverted over a cremation burial which has been placed in an unprotected pit. Again, only 4 such burials are found in the study area, mostly in northern Co. Cork (see O'Kelly and Shee, 1974, 83).

On the basis of pottery evidence therefore, the above mentioned burial types are placed in early/middle Bronze Age times, a period which overlaps with the wedge-tomb building (late Neolithic/early Bronze Age) and the stone circle/alignment building (middle/late Bronze Age) of the extreme southwest of Ireland. Ó Nualláin (1977, 15) points out that since the distribution patterns of the wedge-tombs and the cists are complementary and for the most part mutually exclusive, "the picture is one of two communities living side by side but each preserving its own distinct traditions". However, until the dating of all monuments concerned is more firmly established, such broad generalizations as those put forward by Ó Nualláin should be regarded merely as possibilities rather than established fact.

2.3.2 Settlement and Subsistence. The Bronze Age settlements of W Cork and Kerry do not appear in the archaeological record. There are two possible factors which could explain this unfortunate phenomenon: (i) The masking effect of the blanket peat which has already been mentioned. If one accepts the distribution of burial/ritual monuments in the study area as being representative of the general settlement pattern, then the Bronze Age settlements must lie predominantly in the peaty podzol region, and so are hidden under the rolling bogland. (ii) Probably as a result of factor (i), almost all archaeological investigation/excavation has been concentrated on the more obvious burial/ritual monuments.

There are two possible Bronze Age habitation sites which have been excavated, within the study area. By far the most substantial structure is that at Carrigillihy on the southern Cork coast between the towns of Skibbereen and Glandore (O'Kelly, 1951). This site consists of an oval house, 11 x 7.20 m, constructed of stone-faced walls with an earthen core. This house stood within an oval enclosure (24.6 x 21.5 m), the enclosing wall of which was constructed with stones held in place and given stability by uprights which were set radially. The dating of this house has been rather problematic, since the pottery which O'Kelly regarded as being of Neolithic/early Bronze Age date has since been regarded by Proudfoot (1970, 41) as not being "sufficiently characteristic" to support such a date. The association of a bronze awl with the primary occupation of the house is however, taken by O'Kelly as being sufficient evidence for a Bronze Age date. The only other possible 'house' site from this period is the hut excavated by Fahy (1961) beside the stone circle and boulder dolmen at Bohonagh, Co. Cork. At a distance of c 7 m south of the stone circle, postholes outlining a rectangular area c 7 x 3 m were uncovered, with one central posthole where a roof support could have stood. Because of the structural difference between the two rows of postholes, Fahy (1961, 101) put forward the suggestion that "the southern postholes could have supported a ridge pole and that a second foundation trench, now ploughed out, could have existed some distance to the south." This would have given a house of c 7 x 6 m, not unlike that at Carrigillihy. The absence of any habitation refuse in or around the structure throws some doubt on its interpretation as a house-site and its association with the nearby burial monuments is unclear, all of which adds up to a rather enigmatic situation.

One possible starting point for the future investigation of the settlements of this area, could be the many evidences of field system still to be seen protruding through the peat. Unfortunately these field systems have yet to be surveyed in full and the only excavated stretch has been that at Cashelkeelty (p. 67). Large scale area-excavation of these systems could not only throw some light on the land-use patterns of the area but also the chances are that some of the associated habitation sites would be uncovered within, or close by the systems themselves. Such excavations are at present being successfully carried out on the Neolithic field systems at Behy, Glenulra and Belderg in north Co. Mayo.

Some information about the subsistence economy of the Bronze Age in the study area may also be gleaned from the <u>fulachta fiadh</u> or ancient cooking places which are very common in the south of Ireland, especially in Co. Cork (Ó Drisceoil, 1980). These sites when excavated are usually found to consist of a sunken trough partly surrounded by a mound of burnt stones and with a hearth-site nearby. Water supply is never far distant and at Drombeg (Fahy, 1960) a well had been constructed within the whole complex. On the basis of excavated sites at Ballyvourney and Killeens, Co. Cork, O'Kelly has carried out detailed experiments regarding the mode and efficiency of cooking in such troughs (O'Kelly, 1954). Joints of meat were successfully cooked in a wooden trough containing 100 gallons (455 litres) of water which had been brought to the boil in half an hour. These cooking places are usually considered as seasonal halts of hunting bands, and de Valera (1979, 87) mentions a reference by Keating in his History of Ireland <u>Foras Feasa ar Eirinn</u> to the use of such cooking places by the Fianna in their hunting expeditions during the summer months. It will be interesting to see the conclusions reached by O Drisceoil in his study. The radiocarbon dates for <u>fulachta fiadh</u> range from 1700 bc to 1200 ad (Killeens I and II in Co. Cork have given dates of 1763 $\pm$ 270 bc and 1556 $\pm$ 230 bc respectively) and so they appear to have been in existence in Bronze Age times. In this context it is of interest to note that such cooking places, by their very nature were geared towards a meat-consuming society and the possible significance of this shall be discussed later when considering the agricultural economy of the period.

Finally, a brief consideration of the Bronze Age artefacts which may throw some light on the industrial and technological capabilities of these peoples, information which in turn may be useful when discussing the type of society which had developed in the W Cork and Kerry area. The early Bronze Age has been treated by Harbison (1973) while the late Bronze material has been dealt with by Eogan (1964).

Bronze Age pottery from the study area is extremely scarce. There is one disputed find of Beaker pottery form near Dingle, Co. Kerry and there are of course the above-mentioned foodvessel wares and cinerary urns from the cist and pit burials of Co. Cork. A 'typical' flint assemblage from a Bronze Age site would contain barbed and tanged arrowheads and scrapers of various types. Polished stone axes have continued in use since the Neolithic. By far the greatest number of artefacts however are of copper, bronze or gold.

In spite of the presence of copper deposits and the Mount Gabriel mines (of c 1500 bc) in the study area, in the words of Harbison (1973, 129) "it was the Beaker Folk and the cist builders in the eastern half of Ireland who had most of the wealth and wielded most of the power during the earlier Bronze Age." The majority of metal artefacts from this period have been recovered as chance finds, and therefore this eastern bias of distribution is not the result of excavation/investigation. It is quite possible that most of the actual metalworking and marketing took place some distance from the ore deposits, an idea which is supported by the contents of the Bishopsland, Co. Kildare hoard which is generally considered as the tool-kit of a travelling smith. Some metalworking obviously took place within the study area as evidenced by the presence of at least two stone moulds for the casting of early Bronze Age type artefacts (Harbison, 1973, 133; O'Kelly, 1969). The copper an bronze artefacts found included some thin-butted and flat axes (including the famous Killaha hoard which contained seven thin-butted flat axes, one halberd and one dagger, from near Kenmare, Co. Kerry), a few halberds, socketed spearheads and gold discs.

The later Bronze Age (c 1200-500 bc) saw the introduction of new technologies and ideas, an important aspect of which was the rising popularity of the ornament industry which in Ireland saw its expression in gold objects of startling complexity and beauty. A hoard of gold objects of this period was found at Cappeen, Co. Cork and included three penannular bracelets of differing design, a small plain ring and a rod of rounded cross-section. The majority of bronze artefacts of the study area are placed towards the end of the Bronze Age (Eogan's Dowris Phase) and include bronze horns (Coles, 1967), swords of Eogan's class 4 and perhaps most important are the three hoards, Kilmurry, Co. Kerry and Ballydivlin and Mountrivers, Co. Cork (Fig. 3). The Ballydivlin hoard which contained a socketed axehead and a socketed knife has been classified by Eogan as a craftsman's hoard—possibly that of a woodworker. The Mountrivers hoard contained two socketed axeheads, a copper or bronze bracelet, two gold 'dress-fastners' and eleven amber beads. Finally the Kilmurry hoard contained three penannular bracelets, a spearhead and socketed axehead, all of bronze.

This obvious escalation of metalworking in the later Bronze Age points to an increasing supply of both copper and gold which in the case of the latter, means that considerable trading contacts must have existed between the W Cork/Kerry region and eastern Ireland where the gold deposits occur. The numerous hoards found, suggest that craftsmen were prolific (Eogan, 1964, 325) and the existence of a class of travelling smiths or craftsmen is quite possible.

2.4 Iron Age (c 500 bc–c 450 ad)

The transition from a Bronze Age to an Iron Age mode of subsistence was a gradual process which apparently began around the seventh century bc (Scott, 1979, 195). An understanding of the stimuli which brought about this change has yet to be reached but one should perhaps envisage the arrival of small groups of people who brought with them the Halstatt C artefact designs and most important of all, the knowledge of ironworking techniques. As regards its archaeological record, this period is even more nebulous than its predecessor and instead of being burial-orientated is very definitely 'fort'-orientated—a factor which must reflect a change in social structure throughout the country.

This change can be seen even in the meagre amount of Iron Age remains in the study area. The ubiquitous ringfort, of which there are over 40,000 in Ireland, was being built in early Iron Age times and the tradition continued right up until the Medieval period. This possible wide range in date makes it difficult to carry out any settlement or locational analysis until more extensive excavation may perhaps delimit contemporaneous patterns. Even though the study area contains hundreds of these monuments, less than twenty have been excavated and the indications are that while some such as Garranes and Ballycatteen, Co. Cork may be dated to the fifth and sixth centuries respectively (Ó Riordáin, 1942; Ó Riordáin and Hartnett, 1943), the majority were constructed in the Early Christian period proper and many of the large stone-built examples may even be later. There are however, certain unifying elements within this entire system, such as the basic concept of enclosed settlement, which suggest continuity of tradition for several centuries from Iron Age times.

In spite of the term 'fort' these sites were basically enclosed farmsteads and possessed little or no military significance. The circular enclosures, which have either earthen or stone ramparts, varied from c 15 m to c 60 m in diameter. Within the enclosure were free-standing or lean-to structures, the former being generally round, rectangular or sub-rectangular in shape. According to Ó Corráin (forthcoming) references in early Irish texts mention more than one family living in houses within a single enclosure. The excavated site at Garranes produced evidence of metal and glassworking having been carried on within the fort and according to de Valera (1979, 31) "such manufactories are to be thought of as a degree of self-sufficiency of the homesteads rather than evidence for centralized trading". Buildings such as corn-drying kilns, barns and in the later period, mills, are mentioned in the early Irish laws as lying outside the ringfort enclosure itself.

On the basis of distribution analysis of ringforts in W Cork Fahy (1969) concluded that the inhabitants were "predominantly lowlanders" with 80% of the sites located below 120 m OD and a marked preference was shown for the better Brown Earth and Brown Podzolic soils. A reconstruction of the economy of Iron Age ringfort dwellers in the study area is virtually impossible in the absence of excavated sites which can be firmly assigned to that period. The fort at Garranes which was constructed towards the end of the Iron Age suggests an economy based on mixed farming. Evidence of tillage is present in the form of quern stones, although the possibility of imported grain should not be forgotten.

The animal bones recovered were dominated by those of cattle, followed by pigs and sheep/goat. A surprisingly large percentage (9%) consisted of horse bones which suggest the use of these animals possibly as a food resource but more probably for riding or as draught animals. As already mentioned, both iron smelting and glass manufacture were carried out and foreign trade is evidenced by the presence of imported wheel-thrown pottery of Mediterranean origin (Classes A and B). Whether this general picture of economy can be assigned to the Iron Age as a whole is difficult to say. Mitchell (1976, 158) points out that the pollen diagram from Red Bog, Co. Louth plus others from Limerick, Tipperary, Meath, Antrim and Tyrone, indicate a marked agricultural decline in the Iron Age, especially in comparison with the late Bronze Age activity. This decline he attributes primarily to widespread exhaustion of the soil due to three thousand years of cropping and to the probable introduction of the ard in late Bronze Age times which would have loosened the soils to a greater depth and so encouraged leaching in a period when the climate allegedly deteriorated.

The tradition of the Iron Age hillfort finds little expression in W Cork and Kerry. Raftery (1972; 1976) divides the sixty or so Irish hillforts into three types; (i) univallate types with a general eastern distribution, (ii) multivallate forts with widely spaced ramparts which are found mainly in the western part of the country and (iii) inland promontory forts. Recent excavations at the multivallate site of Rathgall, Co. Wicklow have revealed extensive late Bronze Age occupation of the hilltop and it is possible that it was these peoples who erected the first earthen ramparts (Raftery, 1976, 352). In fact Raftery is now inclined to the belief that the native rather than the exotic contribution was most important for the development of the Irish hillfort and that in may cases, they could simply represent the response of the native population to insecure and politically unstable times. The few excavated examples suggest that permanent domestic habitation was present within the forts but sufficient evidence has yet to be gathered to enable generalisations to be made about the uses to which the enclosed areas were put. The study area contains two examples of hillforts—Cashel Fort, Co. Cork which has an outer rampart of stone and an inner one of earth, and the inland promontory fort at Caherconree, Co. Kerry which consists of a triangular spur defined by almost vertical cliffs on two sides and defended on the third by a wall with terraces on the inner face and a bank of earth and stones (Fig. 3). This latter fort is spectacularly sited at over 600 m OD and encloses an area of c 0.8 hectares. Neither of these forts has been excavated.

The final type of fort which is commonly assigned to the Iron Age is the coastal promontory fort which usually consists of a sea-girt promontory enclosed by means of ramparts across the narrow neck of land. Several such forts may be found along the W Cork and Kerry coasts (Fig.3) and three have been excavated, all in Co. Cork (O'Kelly, 1952). No evidence was recovered to date the construction of these forts and there was little sign of occupation which suggests that these structures served primarily as places of refuge. The occurrence of closely-spaced multivallation in some promontory forts (e.g. Dunbeg, Co. Kerry) has prompted Raftery (1976, 356) to suggest that "they may represent intrusions by individual Iron Age groups in the last century bc possibly from the Venetic area of north-west France arriving via southern Britain". Such intrusions must remain speculation however until dating evidence and evidence of foreign contacts can be established.

Finally, this rather sketchy picture of Iron Age times in W Cork and Kerry is characterized by the absence of crannógs, or lake-dwellings, which occur in other parts of the country (a small island in Inchingeela Lake, Co. Cork may prove to be a crannóg), and burials have not been recorded.

2.5 Early Christian Period (c 450-850 ad)

The proto-historic period of Early Christian times is the final period considered in this study. The date-range of this Early Christian period is rather arbitrary but is generally taken as beginning in the early fifth century with the arrival of the first missionaries and terminating with the arrival of the Vikings in the ninth century. It is not considered necessary to describe in detail the archaeological and historical aspects of this period (see de Paor, 1978; Hughes and Hamlin, 1977) but rather to outline the evidence concerning the organization of society and economy as indicated by the literary sources and excavated sites. The palynological evidence which shall be presented later may then be considered within this context.

The W Cork/Kerry region was one particularly favoured by the early ecclesiastical settlers as evidenced by the many monastic remains concentrated on the Dingle and Iveragh peninsulas (Fig. 3), and the secular settlements are represented by many of the remaining earthen or stone-built ringforts. The Viking raids and settlements in Ireland during the ninth and tenth centuries appear to have made little impact on the remote southwest. The cities of Cork and Limerick are Norse foundations but farther west their activities must have been restricted to raiding incursions which have left little trace on the archaeological record.

2.5.1 The Secular Society. The archaeological record gives a very limited picture of conditions in Early Christian Ireland and the greater part of our information is derived from the many manuscripts of the twelfth century and later which incorporate some material derived from oral tradition dating to the early centuries ad. The majority of manuscripts which survive today were those copied throughout the centuries in the scriptoria of the monasteries and so care should be taken when interpreting the references to pre-Christian traditions because "censoring and alteration of texts have therefore suppressed some of the more openly heathen content" (de Paor, 1978, 66). This unreliability of much of the early literature is beautifully described by the twelfth century scribe who copied out the famous saga of Táin Bó Cuailgne and then declared "But I, who have written this history, or rather fable, am doubtful about many things in this history or fable. For some of them are the figments of demons, some of them poetic imaginings, some true, some not, some for the delight of fools" (translation from Greene, 1945, 88). Inaccuracies in modern translation of the Old and Middle Irish should also be taken into account.

Much of the early Irish literature has recently been examined by O Corráin (forthcoming) with a view to deducing the rural economy of Ireland c 800 ad and the following outline is based on this work. In general the type of society seems to have been tribal, familiar, rural and hierarchical. Kingship was well-established and the kings of provinces lived in large and impressive versions of the ordinary farm settlements. According to Ó Corráin the settlements of petty kings and greater lords were distinguished from those of commoners by a corvée rampart (_drécht gialnai_) but in economy were little different from those of the richer farmers.

The ringforts were the homestead enclosures of the well-to-do farmers with little or no defensive significance attacked. The many souterrains (underground chambers) found within and close by ringforts are generally considered to have served as storage places for foodstuffs and valuables and perhaps for refuge in times of attack. According to _Críth Gablach_ (a law tract of c. 700 ad) the houses within a ringfort ranged in size from 19 to 27 feet (5.8 to 8.2 m)—a dimension which is normally taken as representing the length. The stone-built fort at Leacanabuaile, Co. Kerry contained rectangular houses of similar length. The houses of many of the smaller farmers and peasants were unenclosed structures in the fields.

In the early seventh century, the land was owned and farmed by the _derbfine_, "a four-generation agnatic lineage-group which constituted the family for legal purposes". The land was owned jointly by members of the _derbfine_ and inheritance was partible. It is interesting to note that the _derbfine_ functioned as a joint-farming cooperative and joint ploughing and pasturing were matters for arrangement amongst kinsmen. This arrangement plus the fact that in the case of division of land, utmost care was taken to ensure that each man received his fair share of arable, pasture, good and bad, must have influenced the form of field systems. The arable land was that nearest the farmyards and was generally farmed in strips. Beyond the arable lay fenced pasture-lands some of which were preserved for winter whilst others were used in Spring and Autumn. These pasture-lands may have been dispersed amongst those of other kinsmen. The rough grazing provided by bogland, upland and woodland was also used as commonage by some farmers. Milch cows, usually tended by women were grazed in these pastures and dairy products were stored for the winter.

Towards the end of the seventh century, the complex system based on the _derbfine_ seems to have declined and Ó Corráin considers the plagues of the seventh century as being largely responsible for this change. Smaller land units replaced the larger and the cooperative now consisted of groups of partners of differing status who held farms of unequal size. Partnership had begun to take over some of the functions of kinship.

The ancient law tracts contain much information concerning the practices of cultivation. Ploughing was usually done with a team of four oxen and was restricted to the lands of the more wealthy farmers who pooled their resources for joint ploughing. Cultivation by wooden spade with an iron sheath was the norm amongst the poorer classes. Several sources, including _Crith Gablach_ mention harrowing. Little is known about the manure used except that it was a valuable commodity and usually spread in Autumn. Both ploughing and sowing of cereals were carried out in March and the crops were usually grown in broad ridges which facilitated drainage. The cultivated cereals in order of importance were oats (_Avena sativa_), barley (_Hordeum_), wheat (Triticum) and rye (Secale cereale). Wheat was highly valued and often used by the farmers to pay dues to their chieftains. The work ruadán which was occasionally used in early texts has been translated by Kuno Meyer as 'buckwheat' (Fagopyrum esculentum) but Mitchell (1976, 180) suggests that the term refers to the knotgrass (Polygonaceae) family of seeds in general and that they were grown as a deliberate crop and not just gleaned off the fallows.

The cereals were reaped with sickles and the straw seems to have been cut high up on the stalk. Because of Ireland's heavy rainfall, the cereals had to be artificially dried before milling or threshing and so the kilns were constructed. As with the plough-teams, only the monasteries and better off farmers owned corn-drying kilns. Threshing and winnowing were often carried out within the kiln-houses. The method of storing the grain is not clear, except in the case of the higher class farmers who possessed barns. The grinding of corn was carried out using quern-stones and horizontal-wheel water mills.

Vegetables were also grown, predominantly within the monastic gardens. Onions, leeks, chives and possibly cabbages and parsnips were grown with occasional references being made to peas and beans. Apples were the only fruit grown and orchards were common especially within the monasteries. Finally, flax was an important crop for most farmers, for the production of linen.

In spite of the widespread cultivation carried out during Early Christian times, it was stock-raising and dairying which dominated the agricultural scene—a situation very similar to that of today. Taxes and dues, fines and penalties were calculated in terms of milch cows or fractions of their value while land itself was estimated in terms of the number of cows it could feed. Cattle-raiding was the ultimate act of war. According to the literature, the majority of the cattle were milch cows and only a small number of male animals were raised for breeding and draught. The grazing cattle were watched over by a herdsman during daylight hours and at night were put into a special cattle enclosure. Evidence for winter housing is poor, and the rough grassland on bogs and hillsides and the undergrowth of the woods were specifically reserved as winter grazing.

Milk was of prime importance in its liquid form and also as curd, butter and cheese.

Even though the number of pig bones recovered from excavated sites such as Leacanabuaile, Co. Kerry and Cahercommaun, Co. Clare are minimal (averaging 10%), the pig appears to have been of major importance in the economy of this period. Pig-meat, both fresh and salted was the meat generally eaten by the lower classes and it appears that the animals were slaughtered when quite young.

Sheep were raised principally for their wool, but mutton often formed part of the peasant's render to his lord.

It is difficult to deduce the size of holdings during the Early Christian period but Mitchell (1976, 180) has estimated that a minimum of about 70 acres (c 28 hectares) was needed "if one was to establish a position as an independent farmer". On the other hand, in order to be considered a 'strong' farmer, about 700 acres (c 280 hectares) were needed.

Finally, it is of interest to note that one of the legal tracts lists the penalties for unlawfully interfering with trees and bushes and the severity of the fine was an indication of the value that timber had reached through scarcity. Mitchell (1976, 177) lists the four categories of importance:

NOBLE TREES	COMMONER TREES	LOWER RANKS	BUSHES
Oak	Alder	Blackthorn	Bracken
Hazel	Willow	Elder	Bog-myrtle
Holly	Hawthorn	Spindle	Gorse
Yew	Rowen	Whitebeam	Blackberry
Ash	Birch	Arbutus	Heather
Pine	Elm	Poplar	Broom
Apple	Cherry(?)	Juniper(?)	Rose(?)

According to Fergus Kelly who has been working on these lists, the compiler arrived at the above rankings by taking into account (i) the size of the tree, (ii) quality of its timber, (iii) other uses, particularly of the fruit (Mitchell 1976, 177).

2.5.2 The Monastic Society. The early fifth century saw the introduction of Christianity to the whole of Ireland by wandering missionaries who founded churches under the patronage of bishops and priests. By the beginning of the sixth century however, the Irish social system with its emphasis on family and personal rule was rapidly adopting the concept of the monastic family and its abbot. The foundation of these monasteries was primarily in response to religious aspirations and ideals but before long they adopted ancillary roles as places for fosterage, centres for education and learning and as patrons for the arts. The monasteries dispensed hospitality and to some extent even acted as prisons and so in general, formed an integral part of early Irish society (Hughes and Hamlin, 1977, 15).

The siting of monasteries and related sites within the study area is concentrated very much in the remote western tips of the Dingle and Iveragh peninsulas, with the most dramatic example being the monastic settlement perched on the Great Skellig (Skellig Michael) which consists of a jagged rock rising 715 ft. (c 230 m) out of the sea, off the west coast of Kerry (Fig. 3). These isolated sitings may perhaps indicate houses for ascetic and contemplative communities rather than communities active in educational and social programmes.

The subsistence economy of the monastic settlements was generally similar to that of the secular community which has been described above. Cereal cultivation, dairying and pig farming were all important and most monasteries also kept bees and hens. Fish were an important item on the monastic diet. Both cattle hides and sheep skins were highly valued and used for the making of boats, ropes, shoes, belts, harness, cooking vessels and the vellums on which the books were written. There is a description of the felling of trees in Hisperica Famina which is a seventh century Latin text book probably used in one of the monastic schools (Hughes and Hamlin, 1977, 45). It describes how the trees to be felled were first marked with knives, the roots then chopped through with axes and the trees felled, split with wedges and carried off in carts. Wood was used extensively for domestic furniture, equipment and farm implements.

In conclusion therefore it may be said that the mode of subsistence in Early Christian Ireland was one which depended on a delicate balance of factors. As Ó Corráin (forthcoming) points out, the food supply depended entirely on animal husbandry and cereal-growing with little or no trade in basic foodstuffs to cushion the growing population against shortfalls. Adverse weather conditions frequently caused famines during the eighth, ninth and tenth centuries with resultant fever and plague epidemics and social disorganisation.

CHAPTER 3

A HISTORY OF PALAEOECOLOGICAL RESEARCH IN IRELAND

3.1 Introduction

In this chapter a brief outline shall be made of the palaeoecological research carried out in Ireland since the beginning of this century. This outline shall be restricted to the post-glacial period with particular emphasis placed on the anthropogenic factor in the vegetational history as evidenced in the palynological investigations.

3.2 Palynology

3.2.1 The Study Area. The palaeoecological research which has been carried out in W Cork and Kerry is dominated by that of two workers, Professor Knud Jessen and Professor G F Mitchell. The first scientific programme of investigation was carried out by Jessen in 1934 and 1935, at the invitation of the Committee for Quaternary Research in Ireland. The aim of this programme was to investigate a series of Irish peat bogs in order to advance the studies of late-glacial and post-glacial development of flora and climate, subjects which had previously been touched upon by writers such as Praeger (1932) and Erdtman (1924, 1927, 1928). The results are described in detail in Jessen (1949, 85-290).

The majority of sites included in Jessen's study were the northern and western parts of the country and a total of 10 in the study area were examined (Fig. 4). The choice of sites was influenced by the presence of interesting stratigraphy or the finding of datable archaeological artefacts during peat-cutting. Pollen analysis was the technique usually employed and in spite of severe limitations such as widely-spaced samples, low pollen counts and the absence of an independant dating method such as C^{14}, the outline zonation scheme for post-glacial vegetational succession arived at by Jessen is still referred to today (Fig. 5).

The post-glacial sequence opens with Jessen's Zone IV — the Pre-Boreal of Blytt (1876) and Sernander (1908), which is known from three profiles within the study area, the Long Range blanket bog near Killarney, the nearby river-meadow at Devil's Bit and Emlaghlea blanket bog just west of Waterville. (Fig. 4) Stands of birch and poplar trees were spreading but *Empetrum* (crowberry) heaths were still widespread. Pine values were low and Jessen concluded that pine may not have become an important woodland species until Boreal times.

Zone V is represented at The Long Range and Emlaghlea and the lower border of this zone was placed at the rational border for hazel. Hazel and birch are the dominant trees, which appear to have covered the lowlands and lower mountain slopes.

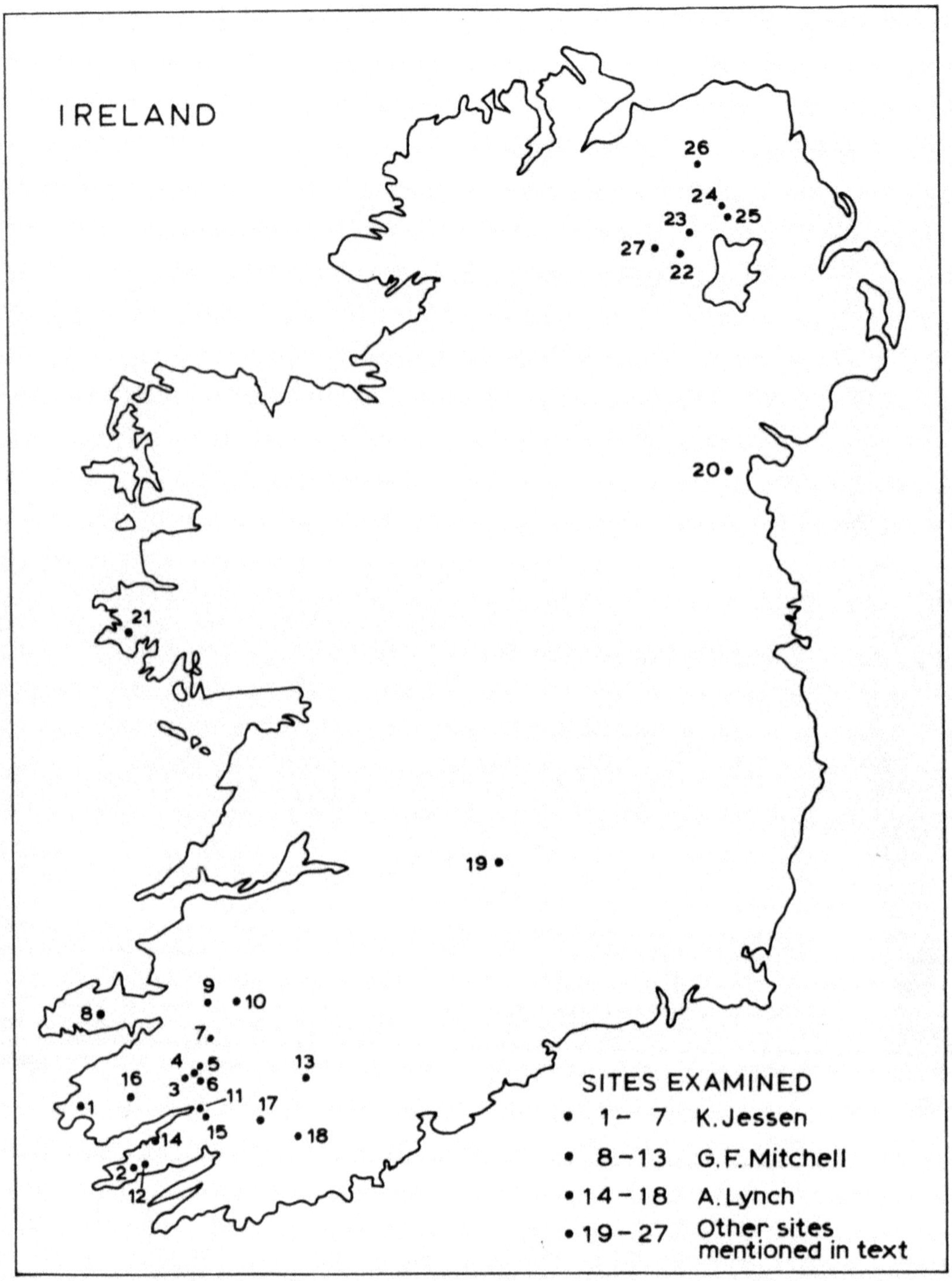

Fig. 4 Locations of palaeobotanical investigations in the study area and others referred to in the text. 1 Emlaghlea. 2 Tooreen More West. 3 Galway's Bridge. 4 Long Range. 5 Devil's Bit. 6 Crinnagh River. 7 Knockasarnet. 8 Coumanare. 9 Ballinvarscal. 10 Cordal. 11 Milleens. 12 Gortagenerick. 13 Milmorane. 14 Cashelkeelty. 15 Dromatouk. 16 Dromteewakeen. 17 Maughanasilly. 18 Cullenagh. 19 Littleton. 20 Red Bog. 21 Dolan. 22 Ballynagilly. 23 Slieve Gallion. 24 Newferry. 25 Ballyscullion. 26 Fallahogy. 27 Beaghmore.

Zone VI which begins at the rational border for elm and ends at the rational border for alder is present again in The Longe Range and Emlaghlea profiles but also appears in that from the blanket bog at Tooreen More West at the tip of the Beara peninsula. Jessen divides this zone into three sub-divisions: (a) hazel reaches a maximum with rather high pine and elm values, (b) elm reaches a maximum in the northeastern part of the country but little elm is present in the southwest, and hazel values fall in the southwest, (c) pine increases dramatically often to reach a post-glacial maximum.

The Atlantic period is termed Zone VIIa by Jessen and is characterized in the southwest by rather high pine, oak and alder values, and in the northern part of the country elm maintains relatively high values. Ilex and Hedera (ivy) are spreading and both blanket bogs and raised bogs are beginning to form. This zone was recognized in five profiles from the study area, Togherbane, Knockasarnet, The Long Range, Emlaghlea and Tooreen More West.

Zone VIIb (corresponding with Sub-Boreal times) opens with a marked decline in elm pollen values and Jessen places its upper boundary at Recurrence Surface C (Grenz-horizont III of Europe) which he considers to be "contemporaneous in those bogs in which it occurs" (1949, 250). This is a zone in which elm declines and pine almost disappears completely from the pollen record. Oak increases rapidly as do alder, hazel, holly and ivy. The apparent decline in forest area and the appearance of Plantago lanceolata (plantain) is attributed by Jessen to the activity of man and the opening up of pasture land. Myrica gale appears on the bogs of W Ireland for the first time. The presence of this zone was recognized in all the profiles mentioned for the previous zone with the exception of Knockasarnet.

Jessen includes the Sub-Atlantic and Historic Times together in Zone VIII and deals with them quite superficially. A climatic deterioriation during Sub-Atlantic times is deduced from the increased growth-rate of the blanket and raised bogs and Jessen also attributes much of the woodland disappearance of this period to increased rainfall and possible increase in wind (Jessen, 1949, 263). Man's activities must also have played a part in the deforestation and Jessen concluded that "it is probable that as far as the Irish woods were destroyed by man this was primarily due to grazing of cattle which checked the renewal of woods and the clearance of fields probably took place in a gradual manner." (Jessen, 1949, 265). No evidence was found for 'occupation phases' or landnam as described by Iversen for the prehistoric period in Denmark.

Jessen's work was continued by G F Mitchell in the following decades when palynological investigations were carried out all over Ireland but especially in those areas not previously covered. In one of his early papers, Mitchell attempted the dating of archaeological objects to certain zones (Mitchell, 1945). These artefacts which ranged from bronze daggers to amber beads were usually found during peat-cutting operations and their findspots subsequently reported which enabled samples for pollen analysis to be taken. The study area yielded several such finds: a hoard of three flat copper axes from Cordal, a bronze dagger from Glannalappa East and a bronze rapier from Togherbane, all from Co. Kerry and all dated to the upper part of Zone VII (early/middle Bronze Age); a bronze cauldron from Ballinavariscal, Co.

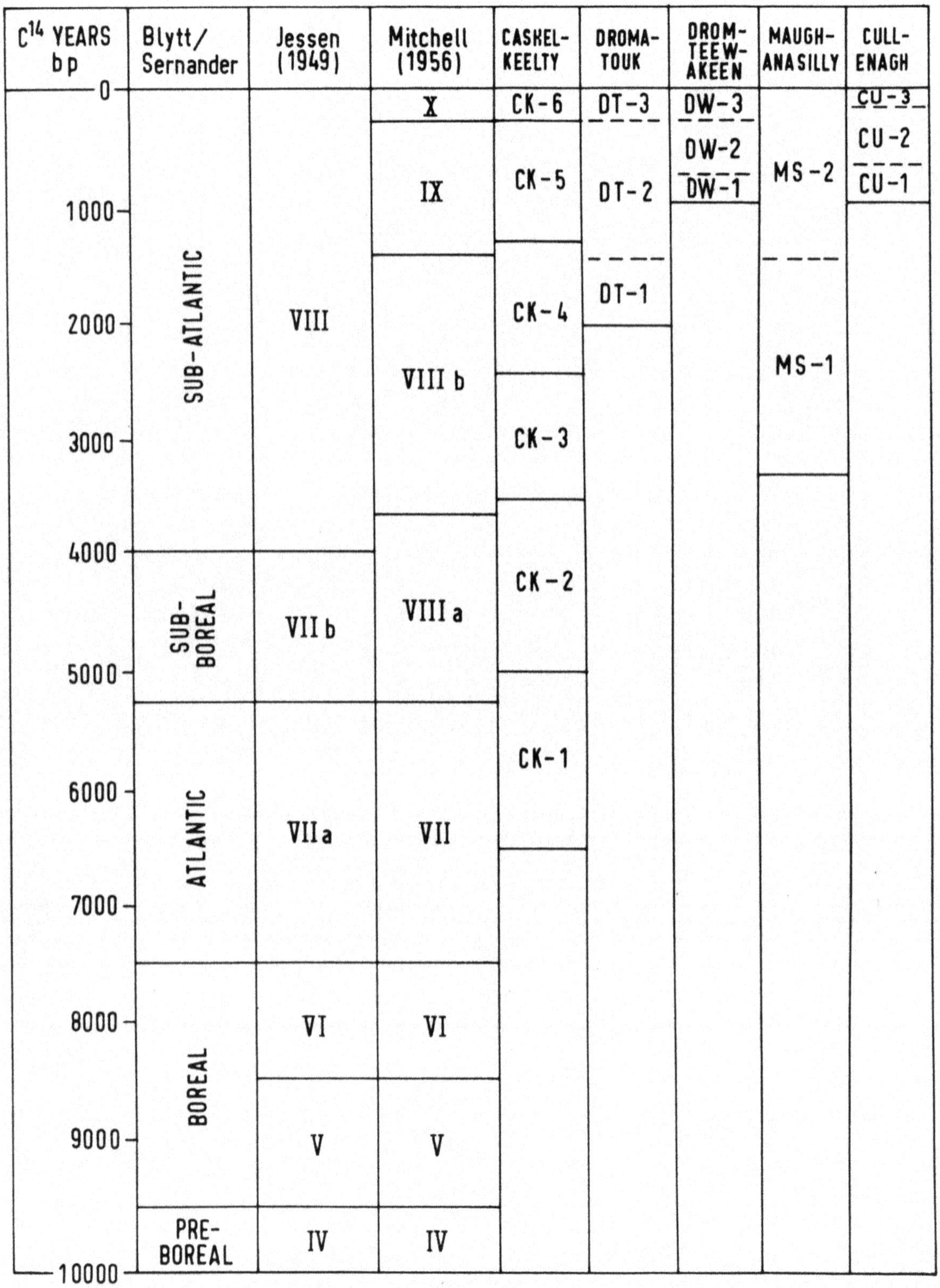

Fig. 5 Summary diagram showing correlation of various zonation schemes referred to in the text.

Kerry and dated to Zone VIII (late Bronze Age); amber beads and a glass bead with a gold mount from Milmorane, a wooden causeway with fragments of a bronze cauldron from Coorycommane and a box of alder wood from Gortagemerick, all in Co. Cork and dated to Zone VIII.

Mitchell's criteria for establishing the Zone VIc and VIIa transition were slightly different from those of Jessen, in that instead of accepting the rational border for alder, he places the boundary where alder is rising steadily, hazel has fallen back to a minimum and pine is also decreasing (Mitchell, 1951, 113). Otherwise his zonation in this early paper (1951) follows that of Jessen. Eight further sites were investigated in W Cork and Kerry (Fig. 4) but again the results are of limited value because pollen counts were low (c 150 AP excluding hazel), intervals between samples were large (c 10 cm) and the pollen associated with agricultural activity were not recognized due to war-time lack of communication with Iversen's work in Denmark. What does emerge however from these investigations is that pine played a dominant role in the woodland of SW Ireland in post-glacial times until zone VIIb when it appears to have been replaced by alder, a change which Mitchell suggests may have been due to increasing oceanic conditions. Considerable emphasis is placed on equating the Recurrence Surface (RS C) of Ireland, which was traced in many of the lowland raised bogs, with that of Scandinavia (RY III) and so placing it at c 500 bc.

Finally, there is the work of Welten (1952) who investigated two high-altitude blanket bog sites in the Macgillycuddy Reeks and concluded that the continuous presence of pine during zone VIII in both diagrams must indicate that it did not in fact disappear entirely in all parts of Ireland after zone VII but survived, perhaps only in small stands, in certain refuges until wide-scale planting took place in historic times.

From the above account it can be seen that the study area has received little attention in recent decades from palaeoecologists and no environmental research has been carried out within an archaeological context.

3.2.2 <u>The Remainder of Ireland.</u> Central and Northern Ireland have fared much better in terms of more scientific investigation of early man's adaptation to his environment. Mitchell continued with his research, concentrating on the raised bogs of central and eastern Ireland (Mitchell, 1956) and his reconstruction of man/land relations through the centuries is perhaps best illustrated by the investigations at Littleton Bog, Co. Tipperary (Mitchell, 1965) and at Redbog, Co. Louth (Mitchell, 1976). Radiocarbon determinations for the Redbog profile have led Mitchell to the belief that "the major agricultural episodes can be traced throughout Ireland and are essentially synchronous" (Mitchell, 1976, 134). The zonation has now changed slightly from that of Jessen's and is based on changes in the pollen record brought about by human activity rather than climatic variation (see Fig. 5 for correlation of different zonation schemes). The first sign of disturbance in the natural vegetation is in Neolithic times (c 3300 bc-c 1800 bc) when it is postulated that groups of farmers drifted slowly through the countryside making clearances by killing off stands of elm trees and abandoning their cultivation patches after a usage period of anything up to two centuries. The abandoned clearings were usually invaded by secondary woodland consisting of much elm and ash. An intensification of agriculture begins at c 1800 bc which Mitchell attributes to the arrival of more technically advanced middle Bronze Age peoples. The secondary woodland was cleared and the cultivated area in general was extended but according to Mitchell, still restricted to the better soils. In later Bronze Age times (from c 1400 bc) a further intensification in farming activity is suggested with the cultivation of an increased range of crops suggested by the appearance of many new species of weeds including Artemisia (mugwort) and the Chenopodiaceae (goosefoot). A migratory type of agriculture is put forward as an explanation for the alternate expansion and contraction of the cleared areas. A decline in Quercus values suggests that the heavier soils are now being cleared and cultivated.

A decline in agricultural activity c 200 bc is evidenced by an increase in arboreal pollen and a reduction in herb and weed pollen. This Iron Age lull in activity is difficult to explain but as Mitchell (1965, 128) tentatively points out, the late Bronze Age if left to itself, may have drifted into "economic stagnation". About 300 ad however, there is an abrupt renewal of woodland clearance, weeds and herbs reappear and cereal pollen is continuously present. In the following centuries the clearances extend, a factor which is usually attributed to the spread of monasticism with its attendant cultivation and husbandry techniques. It is possible that it was the introduction of the more efficient mouldboard plough which facilitated such an expansion of tillage. This phase of activity ended c 1000 ad when another lull is evident and in the following centuries there occur many fluctuations in the intensity of agricultural activity which may be correlated with various historical episodes such as the Viking, Norman and English invasions and settlements.

Palynological investigations combined with detailed radiocarbon dating have been carried out in Northern Ireland over the past decade, resulting in the formulation of several hypotheses concerning the form and duration of prehistoric land clearance and agricultural activities (Morrison, 1959; Pilcher, 1969, 1970, 1973; Pilcher and Smith, 1979; Pilcher, Smith, Pearson and Crowder, 1971; Smith, 1961, 1964, 1970, 1975). The first recognition of Neolithic farming activity which followed the classic Iversen landnam sequence of woodland clearance, cultivation and woodland regeneration, was made at Parkmore by Morrison (1959) and ever since a similar sequence has been outlined in almost every pollen diagram covering that period. At the well-dated site of Fallahogy, Co. Derry (Smith and Willis, 1961-2) the woodland clearance was carried out between 3380 ± 120 (Q-555) and 3240 ± 120 bc (Q-653), the farming activity reached its height at 3340 ± 120 bc (Q-556) and the woodland regeneration commenced at 3310 ± 120 bc (Q-557). The forest regeneration appears to have been complete by 2910 ± 120 bc (Q-654) and agricultural activity was not renewed again until c 2500 bc. The regeneration of the forest in the third stage of the landnam phase leads Smith and Willis to suggest that the agriculture was on a shifting basis, but no evidence was present to indicate whether arable, pastoral or mixed farming predominated. The Fallahogy woodland clearance takes place slightly earlier than that at other Northern Ireland sites, e.g. Cannons Lough, Co. Derry (Smith, 1961) and Slieve Gallion, Co. Tyrone (Pilcher, 1973) where dates of c 3000 bc have been obtained.

The more recent work by Pilcher at Beaghmore, Co. Tyrone and Ballyscullion, Co. Antrim (1971) has succeeded in further subdividing the agricultural stage of the landnam phase into a possible arable period lasting c 100 years followed by a period of 150-200 years when pastoral farming more than likely predominated before woodland regeneration began in earnest. The investigation of the Neolithic and Bronze Age settlement at Ballynagilly has produced an even more detailed subdivision of agricultural phases (Pilcher and Smith, 1979). It is possible that human activity is registering on the pollen record as early as c 3600 bc but the first definite clearance takes place at c 3250 bc and is followed by two phases of early Neolithic activity (the first possibly arable and the second pastoral) separated by a slight woodland regeneration. By 2500 bc human activity is greatly reduced and fairly dense secondary forest is established. Late Neolithic farming is poorly represented and the next real activity phase is the Beaker clearance of secondary forest at c 1800 bc and the establishment of a mixed farming economy. Gradually throughout the early Bronze Age, the spread of heathland conditions caused a change-over to predominantly pastoral farming before the more extensive forest clearances of the later Bronze Age. It is of particular interest that the oak-plank wall of a rectangular house at Ballynagilly has given a date of 3215 ± 50 bc (UB-201) which is indistinguishable from that of 3270 bc for the beginning of woodland clearance and so suggesting a temporal association. Woodland management in the form of pollarding of oak and maybe hazel is tentatively suggested for the late Neolithic (2220-1800 bc).

In all the sites mentioned above, the beginning of Neolithic clearances is evidenced by a sharp decline in elm values. The explanation of this ubiquitous elm decline has been the subject of much discussion and argument. The more recently favoured explanation is the felling of pure stands of elm which would have occupied the richer soils of an area (Smith, 1964, 1970; Pennington, 1975; Sims, 1973). At Ballynagilly however, Pilcher and Smith postulate that the decline in elm values may have been caused by the use of elm bark as a food source for the settlers themselves, or possibly by the stripping of bark from the trees by cattle grazing in the woodland during the winter months. The actual clearing of the settlement area on the other hand, involved the burning off of a mixed forest assemblage including oak, hazel and pine (Pilcher and Smith, 1979, 359).

Even though the majority of detailed investigations has been carried out on lowland sites, the work of Kirk (1973-1974) and Goddard (1971) has shown that agricultural activity in the form of cereal cultivation also took place at altitudes of up to 1500 ft, (457 m) OD during Neolithic times in Northern Ireland. Goddard (1971) has also examined the dating and possible causes for blanket bog initiation in Northern Ireland. Her conclusions state that while climatic deterioration must have played an important role, it was the activities of man especially in the form of woodland destruction through burning and cattle grazing, which were the decisive factors in the initiation of blanket peat accumulation. The effects of human interference became evident in Bronze Age times and the initiation of peat growth centred on three different dates in separate areas, 1800 ± 200 bc, 1200 ± 100 bc and 800 ± 100 bc.

The west of Ireland has received little attention in terms of palaeoecological investigation. The Department of Geobotany, University of Nijmegen, the Netherlands is carrying out a detailed investigation of the recent vegetation of SW Connemara (Beckers, Brock and Klerkx, 1976) and in connection with this research two pollen diagrams from blanket bog deposits at Roundstone and Dolan, Co. Galway were prepared (Teunissen and Teunissen-van Oorschot, 1980). Unfortunately insufficient samples were taken for C^{14} dating but it is of interest to note that cereal-type pollen grains were identified in the later half of the Atlantic period (Jessen's Zone VIIa) c 3500 bc. It shall be of interest to hear the results of work currently being undertaken by G Burenhult at the passage-tomb cemetery of Carrowmore, Co. Sligo and that of Dr M O'Connell concerning the environmental background of the pre-peat field-fence systems in Co. Mayo.

3.3 Seed Analysis

An indication of the status of seed analysis within an archaeological context in Ireland is that the standard reference is still the work of Jessen and Helbaek in the 1940s (Jessen and Helbaek, 1944). Little attention has been paid to the systematic retrieval of macro-botanical remains during excavation and therefore our knowledge of the range of plants used or details of crop husbandry during the prehistoric period in Ireland, is extremely limited.

Jessen and Helbaek's work involved the examination of carbonised seed remains and seed impressions in pottery from prehistoric and early historic times in Great Britain and Ireland. The Neolithic pottery from Ireland produced impressions of spikelets belonging to einkorn (*Triticum monococcum*), emmer (*Triticum dicoccum*) and *Triticum* sp., — all from Dunlay cairn, Co. Antrim and the impression of a grain of naked barley (*Hordeum* var. *nudum*) was identified from Whitepark, Co. Antrim. An impression of a charlock seed (*Sinapis arvensis*) was found at Portsteward, Co. Antrim.

The early Bronze Age pottery produced no impressions but 12 impressions of naked barley (*Hordeum* var. *nudum*) seeds from Cos. Wicklow, Kildare, Down, Antrim, Derry and Tyrone may be assigned to the middle Bronze Age. The late Bronze Age impressions include 20 naked barley (*Hordeum* var. *nudum*) seeds and two hulled barley (*Hordeum vulgare*) fruits from Cos. Carlow, Dublin and Antrim, and one fruit of the weed pale persicaria (*Polygonum lapathifolium*) from Gorey, Co. Wexford. A late Bronze Age impression of a flax seed (*Linum usitatissimum*) was identified from Agfarell, Co. Dublin.

Iron Age seed remains were not found but many carbonised flax seeds (*Linum usitatissimum*) were recovered from a Viking context at Lough Gur, Co. Limerick, together with seven carbonised seeds of corn-spurrey (*Spergula arvensis*). An early Christian impression of a naked barley grain (*Hordeum* var. *nudum*) was found at Kilmoyle, Co. Antrim.

Bearing in mind the inherent bias in the extremely small size of the sample, the results provided by Jessen and Helbaek's investigations suggest that wheat (*Triticum*) was the dominant cereal of late Neolithic times in Ireland, to be replaced during Bronze Age times by naked barley (*Hordeum* var. *nudum*). Flax was desinitely grown during Viking times.

3.4 Archaeozoology

The prehistory of the W Cork/Kerry region has received little attention from archaeozoologists, largely as a result of the extremely poor preservation of bone material in the acidic peaty podzol soils. Much of our information regarding the animal husbandry and hunting of the prehistoric period shall therefore have to be based on the bone material recovered during excavation of sites throughout the remainder of Ireland.

The early insulation of Ireland (probably during late glacial times) which must have hindered the immigration of animals has resulted in a restricted fauna in Ireland (as with the flora which has already been mentioned above p. 7). Few species are definitely known to have been native to Ireland and the economy of the Mesolithic communities who occupied the northern half of the country seems to have been based on fishing, the collection of molluscs and crustaceans and hunting of birds and sea mammals such as seals, with the hunting of terrestrial animals such as red deer, playing a secondary role (Woodman, 1978, 362).

The majority of Neolithic, Bronze and Iron Age sites which produced bone material, were excavated during the early decades of this century and so in most cases the species identification is the only information given and rarely are the actual numbers of bone fragments quoted or the minimum number of individuals calculated. A summary of this information is given in van Wijngaarden-Bakker (1974).

The process of replacement of the Mesolithic economy by a Neolithic one based on animal husbandry and plant cultivation is little understood, but it is generally accepted that the arrival of Neolithic communities familiar with these new techniques and bringing domesticated stock with them, must be postulated in order to explain some of the changes. The bones from the house sites at Knockadoon and the stone circle at Grange, Lough Gur, Co. Limerick indicate an economy based 95-99% on cattle herding. Pigs and sheep/goat constitute only 5% of the bones of the domestic animals and a small number of dog bones were found at all the Lough Gur sites. Horse bones were also found in late Neolithic/early Bronze contexts at several Lough Gur sites. Red deer and a wide variety of birds were most often hunted while brown bear, hare and wild cat were less often the subject of chase. Fishing must have formed an important aspect of the economy of coastal, riverine and lakeside Neolithic settlements but since sieving techniques were not employed during these early excavations, fish bones were rarely recovered.

The early Bronze Age economy is perhaps best illustrated by van Wijngaarden-Bakker's work on the bone material from the Beaker settlement c 2000 bc) at Newgrange, Co. Meath (van Wijngaarden-Bakker, 1974; and forthcoming). A total of 98% of the bones belonged to domestic animals with cattle and pigs being the primary food animals. Sheep/goat appeared to play only a minor part in the economy and the remains of several dogs were identified. Of considerable interest is the presence of domesticated horse which it is postulated, was introduced by the Beaker people (van Wijngaarden-Bakker, 1974, 347). The horse was probably used for traction or transport rather than food. The wild animals found comprise red deer, mountain hare, wild boar, and wild cat. Work on the earlier Beaker material is now nearing completion and the results provide additional information about the Newgrange animal husbandry (van Wijngaarden-Bakker, forthcoming). The cattle appear to have been bred for the production of carcasses of three to four year olds and fat production was also an important aspect of the cattle husbandry system. Some of the steers may have been used as draught animals and it is also suggested that cows were not kept specifically for milk production, and that breeding for secondary products was not carried out. Sheep must have been kept primarily for their wool.

The Iron Age sites which have produced animal bones include one promontory fort, hill-forts and ringforts. Again, cattle, pig, sheep/goat and horse are present and game consisted of red deer, fox, marten, badger and hare. The percentage of sites yielding red deer bones is less than during the Bronze Age which may mean that due to the gradual deforestation of the country, hunting was declining in importance.

Finally, the Early Christian cranńogs of Lagore and Ballinderry I suggest a general decline in cattle rearing with an accompanying increase in pig rearing but such trends may have been due to purely local circumstances rather than to a general change in economy. Unfortunately the remaining Early Christian cranńog sites were insufficiently recorded to further test these hypotheses.

3.5 Summary

On the basis of palaeoecological investigations carried out to date, the following summary may be made of environmental conditions throughout the prehistoric period in Ireland in general:

(i) Mesolithic (c 7000-c 4000 bc). Extensive woodland coverage consisting of *Corylus*, *Pinus*, *Quercus*, and *Ulmus*. Human activity was apparently restricted to the northern half of the country (i.e. on the basis of present evidence) and the economy was based primarily on fishing and trapping of small game and birds, with hunting playing a secondary role.

(ii) Neolithic (c 4000-c 2000 bc). *Ulmus* declines and *Alnus* increases and the woodlands in general are being subjected to clearances for settlement and cultivation. Both raised and blanket bogs are beginning to form. Agricultural activity appears to have taken the form of woodland clearance, cultivation/grazing followed by forest regeneration. Cereals grown included *Triticum* sp. and *Hordum* sp. and the animal husbandry was dominated by cattle herding. Fishing must have been of importance and red deer and birds were regularly hunted.

(iii) Bronze Age (c 2000-c 500 bc). General decline in woodland, *Pinus* virtually disappears and *Corylus* and *Alnus* expand. Grassland increases and intense agricultural activity is evidenced by the appearance of a wide range of weed species. The ard was probably used in ploughing and *Hordeum* sp. seems to have been the dominant crop. Cattle are still of major importance, both for their meat and their fat and pig-rearing is well established. Sheep were kept primarily for their wool and the domesticated horse is present.

(iv) Iron Age (c 500 bc–c 450 ad). Some regeneration of woodland, particularly of *Corylus* and *Fraxinus*. Slight decline in grassland and weeds of cultivation which suggests an apparent lull in agricultural activity. Cattle, pig, sheep/goat and horse are present but hunting appears to have declined. Bogland expands rapidly.

(v) Early Christian (c 450–c 850 ad). Woodland clearance is renewed and grassland again expands. Possible introduction of the mouldboard plough by the monastic settlers may account for the intensification of tillage. Mixed farming was the rule.

CHAPTER 4

AIMS AND METHODS OF THIS STUDY

4.1 Aims

The archaeology of today is largely concerned with studies of sociocultural systems both ancient and modern, which comprise a complex network of subsystems such as the social, psychological, linguistic, religious, economic and material attributes of a given community. A study of any one of these subsystems can only be meaningful if their interdependence is recognized and the various attributes are considered within the context of the "sociocultural and environmental system precipitating them in that particular mutually adjusted configuration" (Clarke, 1978, xvii). The relationship between a cultural system and its environmental system is a close one and is such that fluctuations in the latter may bring about changes in the cultural system as it attempts to adapt. Clarke's external environment of culture systems includes not only flora, fauna, geology and climate, but also other cultural systems. In this study however, environment is defined as 'natural environment' and does not include external sociocultural systems. In its broadest sense, the aim of this study is to try to reconstruct the environmental system(s) of SW Ireland during prehistoric times. Such a reconstruction should be of considerable use in future studies of sociocultural systems as a whole in this area. The secondary aim of this study is a consideration of some attributes of one particular cultural system, i.e. of the stone circle/alignment builders, within its environmental setting.

According to Butzer (1964, 337) the study of man-land relationships in prehistory may be directed towards three major goals:

(i) Understanding the regional environment which includes climate, vegetation and soils;

(ii) Understanding the regional food resource base or 'economic area';

(iii) Understanding the local setting of a site, i. e. location with respect to the terrain, hydrography etc.

A complete reconstruction of the environmental system of any particular period is rarely possible but it is important to understand the degree of complexity of the system and thereby recognize the extent to which the economic activities of any particular community were restricted by the external environment. Of special interest is the possibility of attributing changes in, or differences between the economic activities of communities, to differing environmental settings. The more specific aims of this study may therefore be considered within the above-mentioned framework:

Regional Environment	(a) To reconstruct the post-glacial vegetational succession of the W Cork and Kerry region; (b) To reconstruct the prehistoric soil conditions of the W Cork and Kerry region;
The 'Economic Area'	(c) To date the initiation of blanket-peat growth in the W Cork and Kerry region and consider the extent to which man's activities were responsible for this initiation; (d) On the basis of the results of a, b and c above to try to assess the relationship between man and his environment in the study area throughout the prehistoric period;
The Local Setting	(e) To examine some stone circle/alignment sites at a more local level in order to gain some information about factors which influenced choice of settlement site, factors which would have impeded or facilitated human movements and about the sociocultural system of the communities in general, including the time element involved.

4.2 Methods

Details of the methods employed in this study are presented here in so far as they are relevant for an objective evaluation of the results.

4.2.1 Palynology. Pollen analysis is the primary tool used for the demonstration and explanation of vegetational changes throughout the prehistoric period in the study area. Many aspects of this technique have been misunderstood, abused and criticized by botanists, geographers and archaeologists alike and as Edwards (1979, 266) points out, "in terms of research into the anthropogenic factor in vegetational and environmental history ..... the palynologist must be clear about the fact that he is using aggregated data which he might hopefully filter out in order to provide a coherent story." The following outline of the problems inherent in pollen analysis itself and in its application to archaeological matters should at least enable the reader to form an objective evaluation of the ultimate conclusions drawn from this study.

4.2.1.1 The Deposits. The deposits used for pollen analysis at all five locations (Fig. 6) are blanket peat together with the underlying boulder clay. Peat deposits have long been favoured for pollen analysis chiefly because they develop in a stratified sequence and hence provide a temporal sequence of events. The temporal development may however be distorted by compaction and possible lateral displacement of the lower peat layers during continued accumulation of the deposit and in some cases inversion within the profile may take place following severe erosion and redeposition. A certain amount of vertical movement of the pollen may take place during the early stages of peat formation but according to Moore and Webb (1978, 10) it is "unlikely that such movement will invalidate the basic palynological assumption of vertical (and therefore temporal) stratification because the subsequent compression of the material within which such vertical transport is possible will greatly reduce the effects of pollen displacement." Erosion and peat-cutting may also cause discontinuities in peat profiles. Even if undisturbed, the unknown rate of sedimentation of a peat profile will introduce problems, since the concentration of pollen within the peat will be a function not only of the initial fall-out rate but also of the rate of deposition and degree of compaction of the peat (Moore and Webb, 1978, 10). Finally, the peat-producing vegetation itself may contribute a large percentage of pollen which is of a local nature which may obscure changes in the more regional pollen assemblages.

A total of eight pollen diagrams were prepared for the study area and of these, only the Cashelkeelty I profile was sufficiently deep (3 m) to introduce the possibility of lateral displacement of peat by the weight of overlying deposits. Such displacement is unlikely to have taken place in this particular case however, since the peat monolith was recovered from a deep but localised hollow (located by probing) in the sandstone bedrock which would have given little leeway for peat movement. The only other disturbance recognized in the peat profiles was in Cashelkeelty IV (Fig. 6) where a cut-away surface is indicated by the presence of a thin layer of small stones and some soil in the peat, and the resulting hiatus in the pollen profile is considered in the interpretation. With the exception of Cashelkeelty I, the rate of sedimentation of the peat profiles has not been worked out in details because of the cost of the essential number of closely-spaced radiocarbon dates. In all profiles however, the basal peat layer has been radiocarbon dated, and where there is no visible change in the degree of humification of the peat-forming vegetation throughout the profile, then a rough estimate may be made of the sedimentation rates. In an attempt to isolate the local elements of the peat-forming vegetation, the macrofossil contents of the long profile, Cashelkeelty I, were analysed at 10 cm intervals (for results see Fig. 17) while for the shorter profiles, any recognizable plant remains were noted during removal of the samples.

The use of pollen analysis on mineral soils is more limited, and two major problems have to be contended with: (i) decomposition and differential decay of pollen grains, (ii) movement of pollen within the soil. The problem of decomposition and differential decay has been discussed in detail by Havinga (1971; 1974) and Dimbleby (1957; 1961). The conclusions reached indicate that pollen will not survive to any great extent in soils whose pH is above 6 and that in acid soils the preservation of the pollen appears to be independent of soil type or soil texture. Certain species however seem to be more resistant to microbial attack than others, which can easily lead to biased conclusions (Havinga, 1971). Where pollen does survive in mineral soil, the vertical movement due to downwash and mixing by soil fauna, further complicates the interpretation of soil pollen analyses. Dimbleby has however concluded that the downwash of pollen through the soil confers on each species a depth distribution pattern characteriestic of its history on the site and even though the pollen at one level may not be entirely contemporary, "broad conclusions about the vegetative history of a site can frequently be drawn" (Dimbleby, 1961, 1). There is however, no intrinsic time sequence in the vertical distribution of pollen in mineral soils. Finally, there is the aspect of pollen analysis of 'fossilized' soils which have been buried during the construction of archaeological monuments. The analysis of such buried soils can reveal important information regarding the vegetation of the locality at the time of construction of the monument.

With the exception of Cashelkeelty I (which rests directly on bedrock), all the profiles analysed for this study include the mineral soil underlying the peat deposits. The results of these mineral soil analyses are represented as histograms in the pollen diagrams (Figs. 18-23) and, in each case, the lowermost sample illustrated is the final one in which any pollen survived. The mineral soil in each case is podzolized boulder clay (see chapter 5) and the depth to which pollen survived varies from 5 cm (Dromatouk I) to 21 cm (Cashelkeelty II) but in the majority of cases, little pollen survived below the top of the illuvial B_2 horizon. The interpretation of these mineral soil analyses shall be considered in chapter 7.

4.2.1.2 Pollen Counts. The process of pollen analysis is one of sequential sampling and sub-sampling which in this instance started with the collection of a monolith from a blanket bog deposit the sediments of which were subsequently sub-sampled at selected intervals. These samples were chemically treated (see section 4.2.3) and a further sub-sample mounted on a microscope slide. The final stage was the counting of a 'representative' sample of grains on a given slide in order to give the true proportions of pollen types at that particular depth. In order to keep sampling errors as small as possible, three slides from each depth were analysed and a minimum of 300 arboreal pollen grains per sample were counted wherever possible. Since this study is aimed at reconstructing the regional environment in all its aspects (and not simply the forest history), the pollen sum used is the total pollen, excluding spores. In the case of Cashelkeelty I however, *Salix* (willow) has also been excluded from the pollen sum because the occurrence of clusters of its pollen grains in the lower samples indicated its local presence and the very high pollen values were obscuring changes in the curves of other species. In many of the analyses, the pollen of Ericaceae, Gramineae and Cyperaceae were eliminated from the counts once a representative sample of each had been counted. This approach was necessary because of the very high proportions of the local bog-elements and even though the pollen counts were reduced there was no significant loss of accuracy in the final proportions.

The palynological aspect of this work is therefore based on the relative frequencies of pollen types, i.e. the frequency of each pollen type is expressed as a proportion of a fixed sum. Certain disadvantages inherent in this method should be recognized however. The most obvious one perhaps is that since these are proportional expressions, a change in one component will produce compensatory changes in all other components; for instance, when the input of local pollen fluctuates for any reason, the proportions of non-local pollen types will also change. Every effort has been made in this study to recognize which changes are real and those which are apparent and in some cases the pollen type suspected of causing apparent changes was excluded from the pollen sum and the diagram drawn again (by computer) in order to check the behaviour of the curve fluctuations. Another problem with proportional representation is that, because the total pollen input is the basis of the relative frequencies, changes in the total input itself will not be recognized in the diagram.

4.2.1.3 Zonation of Pollen Diagrams. In order to facilitate the description and interpretation of the pollen diagrams, those which have sufficiently well-defined changes in the pollen profile have been divided into zones. In the case of Cashelkeelty I, the lower part of the profile had been divided into zones accepting the criteria of Jessen and Mitchell (i.e. for Zones VIIa and VIIb) in order to facilitate comparison, but in the upper part of the profile and in all other profiles from the study area the zones or phases are based on the internal microfossil constituents of the individual profiles and consequently need not have any regional implications (these are referred to as pollen assemblage zones). Each zonation scheme is therefore essentially local in character and is valid only within the locality of the sampled site. It was not considered necessary to employ the elaborate techniques of Gordon and Birks (1972) to establish the zonation schemes (since these techniques simply confirmed the zones established by subjective means) and the criteria adopted will be discussed together with the general details of each profile in chapter 7.

4.2.1.4 Interpretation of Pollen Diagrams. The interpretation of the pollen diagrams may be viewed at two different levels. At the first level, the pollen diagram may be interpreted as a record of vegetational change through time with the inherent assumption that each spectrum of the diagram represents proportionally the contemporary pollen-producing flora which was contributing to the pollen rain. The next stage involves a reconstruction of environment, physical and man-made, on the basis of the established vegetation. This second stage of interpretation is based on the assumption that vegetation does reflect its environment.

The first assumption is concerned with the relationship between vegetation and pollen rain. There are many factors which influence the relative abundance of a particular pollen type at a given level and the reader is referred to Moore and Webb (1978, 97) for a concise evaluation of these influencing factors. Proportional representation poses problems for the interpretation of pollen diagrams and these have been outlined above (p. 45). The land area from which the pollen is derived should be estimated before the area of vegetation represented by a particular pollen assemblage may be established. The study of modern pollen fallout patterns has indicated that under appropriate meteorological conditions, light pollen grains and spores may be carried many thousands of kilometres from their source of origin. Such 'exotic' pollen grains are however usually recognized as such by their very nature (e.g. the occurrence of a non-native *Tilia* grain in Dromatouk I). The study of modern pollen spectra from surface samples has also given useful information about the contribution of local vegetation to the pollen rain (Janssen, 1967, 1970; Goddard, 1971). The general conclusion from these studies is that if long distance and local pollen types (in this case the bogland vegetation) can be recognized, then the remaining pollen count will characterize the overall vegetation type of the locality in question. Goddard's work (1971) showed that the tree pollen percentage in a spectrum gives a good indication of the general aspect of the vegetation from which the sample was taken. Factors such as local topography will also influence the area represented by the fossil pollen spectra and in the present study where all the samples were recovered from tracts of blanket bog in mountainous regions, the pollen spectra seem to represent the local vegetation. An overall picture of the regional vegetation may then be obtained by comparison and correlation of the separate profiles. Other factors which serve to complicate the business of interpreting pollen diagrams are the differences in efficiency of pollen dispersal and pollen production between various species. The wind-pollinated species present in an assemblage are most likely to represent the non-local vegetation whereas entomophilous pollen grains are liable to be of local origin. In an effort to cope with the problem of differential production of pollen, Andersen (1973) suggests that relative production indices may be used to adjust the relative proportion in the pollen diagrams. No such correction factors have been used here however, because a standard correction procedure has yet to be decided upon, but the problem has been kept in mind when interpreting the diagrams. Finally, the reader should also bear in mind factors such as time of flowering of various plants, the physiological conditions under which flowering and pollen production take place and the plant communities which may be tenuously reconstructed from the pollen assemblages.

The second basic assumption underlying the interpretation of pollen diagrams is that vegetation reflects to some extent its environment. Every plant species has an optimum requirement for any given ecological variable and has limits relating to that variable beyond which it will not survive. If any of these limits are approached in a particular habitat, then that limiting factor may be the determinant controlling the presence or absence of the species at a particular site. Even though the environmental requirements of a species may be known, it is usually extremely difficult to determine which component of the overall environment is that which is the limiting factor at a certain time. This problem is particularly clear when dealing with climatic variables, especially since many plants respond rather slowly to climatic changes. In addition to reconstructing climatic changes, it is also possible to determine past soil conditions in an area on the basis of the known ecological requirements of plant species represented in the fossil pollen assemblages.

When dealing with the later stages of the post-glacial period, many of the changes evidenced in the pollen assemblages may be attributed to the activities of man rather than to physical environmental factors. The interpretation of these changes is of vital interest to archaeologists and many of the problems associated with the archaeological application of palynological data are clearly discussed by Edwards (1979). Much of the information concerning prehistoric agricultural activity has been inferred from evidence of clearance activity, which in turn has been inferred from pollen diagrams. As Edwards points out, the examination of such clearance activity is a complex process and one dominated by two major problems—the method of clearance and the purpose of the cleared area. It is seldom possible to deduce the mechanism of woodland clearance since diverse clearance methods such as burning, felling by axe, ring-barking and grazing, all give rise to similar results. It has been generally accepted that the presence of charcoal fragments in a profile at the level of inferred clearance, indicates the burning off of woodland cover but Edwards (1979, 256) maintains that such charcoal "may easily have been derived from domestic fires from nearby settlement, or from the burning of axe-felled trees whose ash was ploughed back into impoverished soil for purposes of enrichment". It is unlikely however that the latter explanation of charcoal occurrence could be applied to the Mesolithic or early Neolithic instances described by Simmons (1969) or Smith (1970b) since this would imply a far more advanced agricultural practise for those periods than present evidence (botanical and archaeological) would suggest. A localised spread of charcoal may indeed represent the remains of a domestic hearth-site but in the absence of domestic refuse and if the charcoal is spread over a wide area (see Maughanasilly p. 70) then woodland clearance by fire (whether intentional or accidental) may be postulated.

Once a clearance episode has been recognized in the pollen record, the nature of agriculture practised within such a clearance is usually considered. Many writers (Turner, 1964; Godwin, 1968; Roberts et al., 1973; Sims, 1978; Pilcher, 1979) have attempted to establish a pastoral/arable ratio on the basis of the composition of the weed flora, but unfortunately there is little agreement amongst the species used by different authors as pastoral and arable indicators. The arable/pastoral index established by Turner (1964, 82 82) on the basis of surface sample analyses is of special interest but is specifically constructed with regard to the environmental conditions of the study areas of Cardiganshire and Shropshire and is therefore of limited value for more widespread use. The problem of differential pollen productivity and dispersal is of particular importance here since the most important arable indicators (i.e. cereals) produce little pollen and are self-pollinating and therefore the number of cereal pollen grains which actually come to be deposited in nearby sediments is very small, while on the other hand, *Plantago lanceolata* and Gramineae which are the most characteriestic plants of pasture, produce pollen in abundance and may easily result in an inferential bias in favour of pastoralism. An additional problem is that our knowledge of the environmental requirements of fossil taxa is based entirely on the requirement of their present-day morphological counterparts and such an assumption may not always be justified.

A rigorous approach has been adopted with regard to the use of pastoral and arable indicators in this study. Only those taxa which may be confidently assigned to either pastoral or arable conditions according to Webb (1977) and Dr A O'Sullivan of An Foras Talúntais, Grassland Nutrition/Ecology Dept., Johnstown Castle Research Centre, Wexford (personal communication) have been used. As a result, the plants accepted as representing stable pasture situations are *Plantago lanceolata*, *Ranunculus* sp. (buttercup), *Succisa pratensis* (devil's bit) and the families Leguminosae and Polygalaceae. Those taken as representing arable conditions are cereals, the genus *Polygonum* and the families Chenopodiaceae, Cruciferae and Caryophyllaceae. The family Gramineae have not been included, because when dealing with blanket peat deposits, the surface vegetation may include many gramineae species (e.g. *Molinia caerulea*) the pollen of which cannot be identified to species level. Even with such a reduced range of indicators however, uncertainties may still arise, for instance *Succisa pratensis* may occur as a bogland plant as well as on upland pasture while Groenman-van Waateringe (1968) has noted that *Plantago* sp. may not necessarily indicate anthropogenic interference with vegetation but can also be derived from ecotonal areas where vegetation communities are unstable in marginal environmental situations. Such limitations as have just been outlined do not however negate the pastoral/arable inference from pollen diagrams but rather serve as a cautionary reminder that such interpretations should be critically considered. An attempt was made to illustrate changing emphasis through time, on pastoral or arable activities by drawing a curve of the indicator species for each site, but in every case, the frequencies were too low to produce satisfactory results (see for example Dromatouk I, Fig. 20).

The spatial distribution of woodland clearances with reference to the sampled site is another problem discussed by Edwards (1979, 257) where the point is made that the necessity for waterlogged deposits or acidic soils has largely restricted palynological research to areas which by their very nature are marginal lands, and hence the inference of nearby prehistoric settlement may not be entirely valid. The distribution of sampled sites of this study corresponds closely with archaeological evidence for prehistoric activity and since the major episodes of soil degradation and peat growth did not commence until around late Bronze Age times, the concept of land marginality may only be applied to such areas for the post-Bronze Age period.

It should be clear from the preceding discussion that the value of palynology as a tool for explaining past vegetation changes and inferring agricultural activities, lies very much in the rigorous interpretation of the data and a clear understanding of the inherent problems of the technique. Where possible of course palynological data should be supplemented by archaeological and other environmental evidence to enable a more complete reconstruction of past man-land relationships.

4.2.2 Selection and Sampling of Sites. The number and distribution of sites sampled was determined by the following factors:

(i) the sampling area itself was determined by the distribution of stone alignments in the W Cork/Kerry region (see Fig. 1) and samples were recovered from tracts of blanket bog in which stone alignments were located;

(ii) those sites most likely to have remained undisturbed by peat-cutting, drainage or deep-ploughing were chosen;

(iii) a representative distribution of sites within the study area was attempted, within the limits set by factor (ii) above;

(iv) a total of eight profiles (from five locations) was considered the maximum which could be analysed and interpreted in the time available;

(v) Cashelkeelty was chosen for excavation since it appeared to provide the opportunity of examining the relationships (temporal and cultural) between stone alignments and stone circles and in addition, the remains of pre-peat field fences suggested a nearby settlement;

(vi) Maughanasilly was selected for excavation as one of the best-preserved examples of stone alignments.

At all sites, samples were removed from peat faces exposed either during excavation or by the digging of pits in those areas not excavated. The peat deposits were removed by means of metal boxes 50 cm x 10 cm x 15 cm, the open sides of which were pushed into the exposed vertical profiles. The samples from the underlying mineral soil were collected using open-ended glass tubes which were subsequently sealed. The metal boxes were sealed in polythene bags and transported to the laboratory where successive 1 cm thick slices were cut out for the palynological analysis. A 2 cm thick slice was removed from the base of each monolith for C^{14} dating and in the case of Cashelkeelty I, a 2 cm thick slice was removed at 10 cm intervals for macrofossil analysis. All such sub-samples were thermally sealed in polythene bags until analysed. When the analysis of the long profile, Cashelkeelty I was completed and the diagram drawn, further 2 cm thick slices were removed from its boxes in order to date the critical levels.

4.2.3 Laboratory Procedure. The treatment and preparation of samples prior to counting was according to the system described by Faegri & Iversen (1975). Acetolysis was used for all samples and those containing clay or sand received additional treatment with hydrofluoric acid. The preparations were mounted in glycerol jelly and the identifications were carried out using the keys and descriptions of Faegri and Iversen (1975, 168-198) and Erdtman, Berglund and Praglowski (1959-61). The identification of cereal-type pollen grains was attempted to genus level using phase-contrast microscopy and the identification criteria of Grohne (1957) and Beug (1961). The size criteria adopted were, diameter of grain $> 40\mu$ and diameter of pore plus annulus $> 7.5\mu$. When comparing glycerol mounted grains with those mounted in silicone oil, a conversion factor of 1.4 was applied to the size of the grains to account for the slight swelling caused by the glycerol jelly.

4.2.4 Seed, Macrofossil and Charcoal Analysis. The seed-sampling strategy of the Cashelkeelty and Maughanasilly excavations was to flotate all the deposits containing carbonised material in an effort to retrive some information about the crop economy and local environment of the period of use of the monuments. Unfortunately the failure to locate the settlement sites associated with the stone alignments and circles meant that the only carbonised deposits recovered were the spreads of charcoal in the upper levels of the leached G horizons and the charcoal incorporated in the stony platform on which the five-stone circle of Cashelkeelty was constructed. These deposits were flotated on site and subsequent analysis showed negative results. The only botanical remains identified were several fungal sporangi (*Cenococcus geophilum*).

The long profile, Cashelkeelty I, was analysed at 10 cm intervals for macrofossil remains which would supplement the palynological data and provide information about the local vegetation. The samples (c 30 ml) were wet-sieved over mesh sizes of 1 mm, .6 mm, 0.25μ and 0.16μ and the residue was examined using a Wild M5 stereoscope. The results are presented in Fig. 17. Even though some information was gained from this analysis it was not considered sufficiently worthwhile to analyse the profile at closer intervals because of the very time-consuming nature of the work.

The charcoal analysed was hand-picked during excavation or removed from the exposed peat and mineral soil profiles at the non-excavated sites. In general only those pieces large enough to provide a clear fracture surface were identified and when several such pieces occurred in one context, five were chosen at random for analysis.

4.2.5 Radiocarbon Dating. Both peat and charcoal were used for dating critical levels in the profiles. The series from Cashelkeelty I were processed at the Radiocarbon Dating Unit of the Palaeoecology Laboratory, Queens University, Belfast while the remainder were processed at the Laboratorium voor Algemene Natuurkunde, Rijksuniversiteit, Groningen. The use of uncorrected radiocarbon dates has been adopted throughout this study but the implications of calibration will be discussed later (p. 127) when the radiocarbon dating evidence in general is being considered.

4.2.6 Computer Techniques. Computer techniques were employed in two aspects of the study, (i) the computation and graphical representation of the palynological data and (ii) the comparison of pollen diagrams.

An ALGOL-60 computer programme compiled by Voorrips (1973, 1974) was used for the computation of the pollen counts and the presentation of the results in tables and graph form. Once the data had been entered on the computer (an IBM 360) the computation of the relative frequencies of species and their graphical representation on the lineprinter could be rapidly obtained. In this way a pollen sum could be altered by the exclusion of particular species and the resultant effects on other curves examined, without the laborious and time-consuming job of re-drawing the entire diagram by hand. For reproduction reasons however, the final published drawings are line-drawings.

The second application of computer techniques arose from the need to compare and correlate separate pollen diagrams and to 'fit' many of the shorter diagrams into the chronological scheme of the long profile Cashelkeelty I. Even though many of the shorter profiles were located up to 40 miles (64 km) from Cashelkeelty, and therefore may have had different local vegetation developments, it was considered useful to attempt a broad comparison of the pollen diagrams using only AP values which most likely represent the regional vegetation type. Those species accepted for comparison were Pinus, Betula, Ulmus, Corylus, Fraxinus (ash), Fagus (beech), Alnus (alder), Ilex, Tilia (lime), Quercus.

The method used was based on a comparison of the curves of the various species in the two pollen diagrams concerned, as opposed to the method proposed by Gordon and Birks (1974) where the spectra are taken as the basic units for comparison. Details of the method are to published shortly by Voorrips (in preparation) and will not be described here. It should however be stated that this technique should only be used as a general guide to suggest possible areas of similarity or difference between pollen diagrams and at all times the results should be combined with evidence provided by independent dating methods such as radiocarbon, and the observations of the investigator. Indeed, the value of the technique lies mainly in its use as an independent check on the similarities noted visually between pollen diagrams.

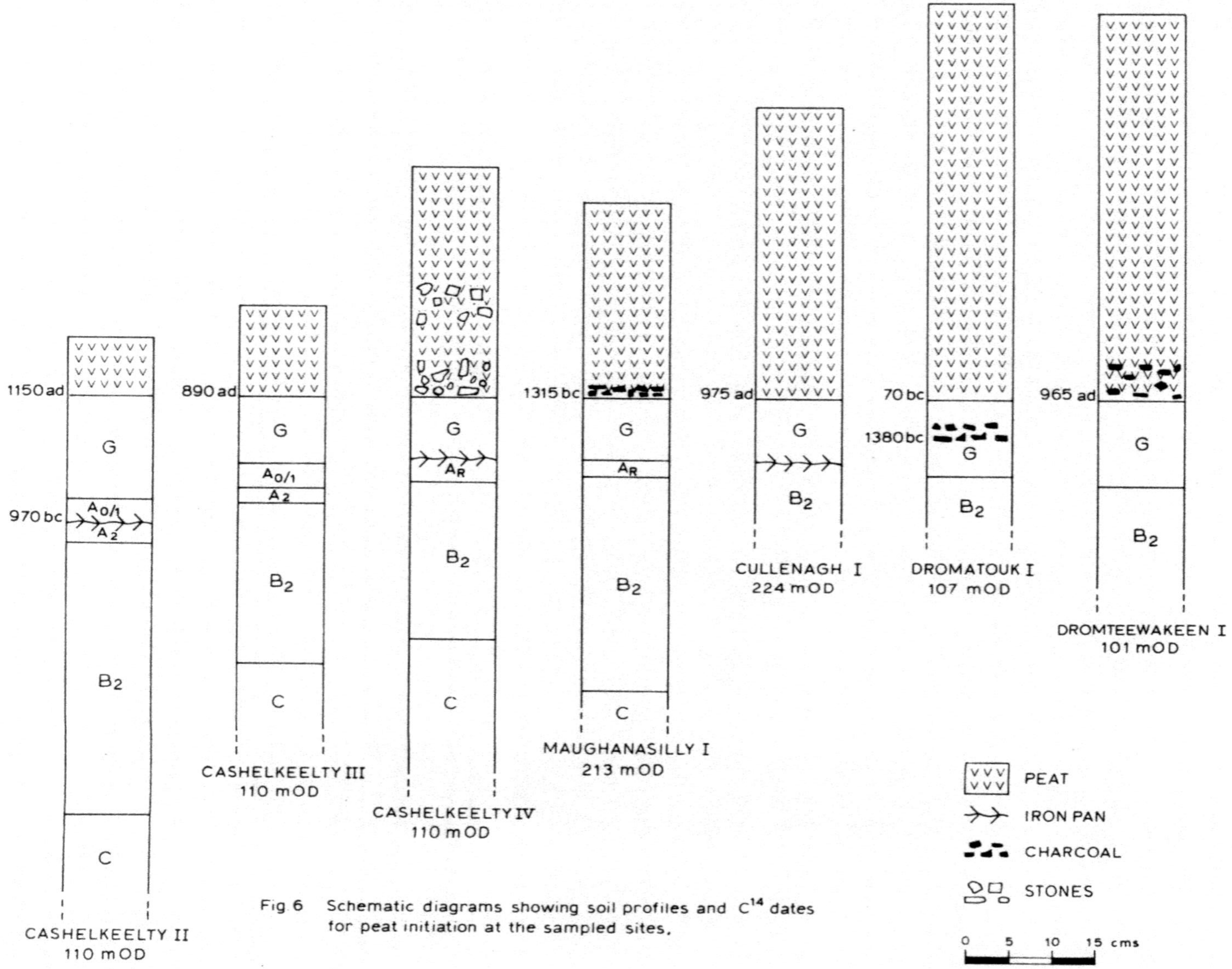

Fig 6 Schematic diagrams showing soil profiles and C^{14} dates for peat initiation at the sampled sites.

CHAPTER 5

SOIL PROFILE DEVELOPMENT AND PEAT INITIATION

5.1 Profile Development

The soils at each sampled site belong to the peaty podzol soil association of the podzol great soil group (Gardiner & Ryan, 1969, 97). These peaty podzols which cover approximately 50% of the study area may be assigned to the 'mountain and hill' physiographic division and while the principal soil is a peaty podzol (75%), climatic peat and lithosols may occur as associated soils (25%). The parent material of the boulder clay at each site is primarily Old Red Sandstone.

Detailed investigations of all the soils within the study area are not yet complete and only those of the W Cork region have been fully surveyed (Conry and Ryan, 1963). The soil horizon notation adopted in this study follows that of Romans (1970). The formation of a classic podzol is generally considered to begin with the accumulation of a surface raw humus mat which is gradually broken down in the A_0, A_1 and A_2 horizons into colloidal humus. The polyphenols produced by this humus attack the uppermost mineral horizon, first weathering the minerals and then forming soluble complexes with the iron and aluminium oxides produced. In free-draining soils derived from acid parent material which occur in areas of high rainfall, the soluble complexes formed will leach down the profile to form the B horizon by progressive deposition of hydrous iron, aluminium oxides and/or colloidal humus. The leached (A_2) horizon, having lost its colouring iron compounds becomes a whitish grey colour and its lower boundary normally fades into the underlying orange B horizon with no clear demarcation. The location of the illuvial B horizons is usually determined by a change in soil density or by the presence of the water table, either of which would slow down or halt the downward movement of metal oxides. Where the situation is such that the mobility of the iron is due to its reduction under conditions of waterlogging, a thin iron pan (B_{fe}) is formed and both Romans (1970, 90) and Limbrey (1975, 139) describe the deposition of humus on top of such pans along the A/B interface.

The soil profiles of the sampled sites (Fig. 6) all reveal a history of podzolization. In each case, a simple podzolization process had taken place initially and where this has been preserved from subsequent alteration, as for example under the old ground surfaces in Cashelkeelty II and Cashelkeelty III, a thin leached horizon (A_2) is visible overlying the illuvial (B_2) horizon which graded into the unmodified boulder clay or C horizon. This means that the podzolization process had begun before the burial of the old ground surface (i. e. the construction of the nearby monuments) and the implications of this shall be discussed later. This primary podzolization can be referred to as a regional podzolization (*sensu* Romans, 1970).

At some time after the initiation and partial development of these podzols the situation was altered drastically. A rise in the water table coupled with intensification of leaching created a continuous hard pan at the level of the new water table. Thin iron pans are visible in the profiles from Cashelkeelty II, Cashelkeelty IV, and Cullenagh I. A similar iron pan was recorded during excavation at Maughanasilly, but it was not a continuous layer and does not appear in the monolith recovered for analysis. The absence of a visible iron pan formation at Cashelkeelty III, Dromatouk I and Dromteewakeen I is not unusual and peaty podzols without a thin iron pan have been widely recorded in Scotland (Romans, 1970, 93) and are termed stagnopodzols by Limbrey (1975, 143).

At Cashelkeelty IV and Maughanasilly I, a thin dark brown/black layer of humus occurs on top of the B_2 horizon. It is unlikely that this represents the deposition of colloidal humus as mentioned by Limbrey (1975, 139) and by Romans (1970, 90) when describing the Humus Podzols of Scotland. Firstly, this humus deposition always occurs on top of both the B_2 horizon of metal oxides and the iron pan where such exists and secondly, such a transported layer of humus should be fine in texture. At Cashelkeelty IV however, a well-developed iron pan lies above the humus layer and in both examples, the humus is of coarse texture. It is therefore possible that these thin humus layers represent the relicts of the A horizon (A_R) of the regional podzolization process which lay below the raised water table level.

After the formation of iron pan which would have impeded the downward percolation of water, the overlying soil became waterlogged and gleying took place with ultimate peat formation on the surface. This gleying process has reduced the upper 7-15 cm of mineral soil to a uniform blue/grey mass of concrete texture (G horizon). Little evidence was found for an impervious black greasy humus layer beneath the peat, as is described by Proudfoot at Goodland, Co. Antrim (1958, 189), and which he suggested was responsible for the waterlogged conditions in which the overlying peat began to form. During the excavations at Cashelkeelty and Maughanasilly, a thin layer of highly decomposed peat was distinguishable at the base of the general peat deposit but since the peat cover was so thin anyway (10-20 cm) the upper mass would normally appear different on the basis of its root content and more oxidised condition.

The various stages in soil formation leading up to the initiation of peat growth, may therefore be summarized as follows:

(i) Degradation of the humus layer begins, and weathering of the minerals in the uppermost soil horizons results in the formation of soluble complexes.

(ii) These complexes are leached downward in the profile and accumulate in the B horizon.

(iii) Water table rises, leaching intensifies with resultant formation of iron pan at the water table level.

(iv) The drainage is impeded by the iron pan and the overlying soil is waterlogged and subsequently gleyed.

(v) Peat formation begins.

These peaty podzol soils are considered to be amongst the poorest in the country and have severe fertility limitations due to deficiencies in phosphorous, manganese and cobalt and to the impeded surface drainage (Conry & Ryan, 1963, A-2b). Land-use is restricted to rough grazing on an extensive rather than intensive scale and in parts, forestry may also be a possibility.

5.2 Factors Influencing Podzol Formation

The whole question of why and when the podzolization process began in the study area is of particular interest when considering the quality of soils available to man throughout the varying stages of prehistory and the impact which prehistoric man's activities had on these soils.

Let us now consider the question of why the podzolization process began in the study area, by taking into account possible influencing factors such as topography, climate and vegetation.

The altitude of sites examined ranged from 101 m above sea-level to 244 m and no significant difference between the soil profiles was seen within this range. This suggests that variations in altitude within this range had little effect on the podzolization process. The topography of the sites is remarkably similar. Even though each site is located at relatively high altitudes in mountainous terrain, the immediate topography is a well-defined level strip of land with steep slopes rising on at least one side providing much run-off water. In the absence of suitable climatic and vegetational conditions however, such topographical details alone cannot account for the onset of degradation in the humus layer.

The role of climate in the development of podzols is inextricably bound up with the time factor. Podzols are known to have been developing in NW Europe since Neolithic times (Proudfoot, 1958, 195; Waterbolk, 1957) right through to the present day which means initiation of the process in various climates since at least the fourth millennium bc. The timing of the development of podzols in the study area is difficult to tie down accurately, and perhaps the best indicators are the soil profiles from Cashelkeelty II and III. It is clear from these profiles that the podzolization process had begun before the old ground surfaces were buried during the construction of the monuments. Charcoal from the old ground surface associated with the construction of the stone circle has given at C^{14} date of 2920 ± 60 bp (GrN-9173) or 970 ± 60 bc. No radiocarbon date was obtained for the construction of the adjacent alignment, but on the basis of stratigraphic evidence, it was constructed before the stone circle, but only by the length of time necessary for the development of a thin humic layer over the disturbed soil (Barber, in preparation, a). The depth of the leached A_2 horizons under the buried ground surfaces was c 2 cm which indicates that the podzolization process was not far advanced before 970 bc. Since the rate of podzolization can vary considerably depending on surface conditions, we can but tentatively suggest that podzolization was beginning in the Cashelkeelty area by c 1000 bc.

Whether or not this date can be applied to the study area in general is not clear. The dates for the basal peat layers in the remaining profiles (see Fig. 6) provide a terminus ante quem for the podzol development at each location, but since these dates range from 1315 bc to 975 ad and the time-lapse between podzol initiation and peat growth is unknown, they do little to solve the problem. At Maughanasilly however, the presence of a gleyed podzol profile beneath a peat deposit whose basal layer contained charcoal dated to 3265 ± 55 bp (GrN-9281) or 1315 bc indicates an earlier initiation of podzolization — perhaps in the first part of the second millennium bc. On the basis of radiocarbon evidence therefore, we may suggest that podzols were developing in the study area since the early second millennium bc but in some localities the development began much later, thereby suggesting that the process is not simply a function of time.

Keeping in mind the time element just established, let us return to the question of climatic influence. The wide date-range within which the podzolization process began, indicates that climatic conditions alone cannot have been the controlling factor. In fact, as has already been mentioned (p. 4) the prevailing temperature level between 4000-1000 bc was mostly 1°-$2^{\circ}C$ above that of today and the cooler, more moist conditions which would have been most favourable for podzolization, did not prevail until after 1000 bc. Similar conclusions regarding the British heathland soils have been reached by Dimbleby (1962, 18) who states that "the onset of podzolization is not attributable to any particular climatic phase of the post-glacial ... "

Finally, the possible influencing role of vegetation cover must be considered. Many writers over the past few decades have discussed the relationship between the removal of woodland cover and soil degradation/podzolization (Cornwall, 1953; Dimbleby 1965; Limbrey 1975) and in some instances, peat growth is postulated as the terminal stage (Moore 1975). Disturbance or removal of woodland cover will seriously disturb the ecological equilibrium on acid soils. The supply of ground water will be increased due to a reduction in the transpiration demand of the vegetation and also a reduction in the water interception of the canopy. The loss of a root system will usually bring about a reduction in stability and base-status of the soil and so physical erosion together with leaching is encouraged. Removal of the tree-cover, especially on sloping ground will also give rise to increased run-off and perhaps waterlogging in hollows. It is very difficult however to determine a definite cause and effect sequence in vegetation and soil changes. The recent work by Andersen (1979) in which he uses analysis of pollen and fungal hyphae fragments in the humus layers and soil profiles of Eldrup Forest, has illustrated quite convincingly that the podzol development was related to the vegetation and mode of humus deposition. Such detailed microfossil studies may in the future help resolve the problem at a local level. For the present study we are restricted to the pollen analysis of the mineral sub-soils and buried ground surfaces to provide some clues as to the vegetation changes which took place immediately before podzolization began. As already mentioned, pollen rarely survives in soils with a pH exceeding 5.5 which means that usually the primary forest is not represented in such pollen diagrams and the earliest vegetation type represented belongs to the period when soil-acidity had become high enough for pollen preservation.

Cashelkeelty III (Fig. 18) provides perhaps the best illustration of conditions prevailing during the early stages of podzolization. The spectrum at depth 25 cm below the surface represents an incipient stage of podzol development which was subsequently sealed by the buried ground surface. There is little indication of the earlier extensive woodland cover established for the locality by the peat profile Cashelkeelty I and extensive soil degradation is suggested by the dominance of Ericaceae (heathers). *Succisa pratensis* which is a common upland pasture or heathland plant, is also plentiful. The only woodland remains are *Corylus* and *Alnus* with a little *Betula*. The picture presented by the old ground surface is very similar, with heather dominant, some hazel scrub and cereal cultivation nearby is indicated by the three cereal type pollen grains. The NAP/AP percentage (standard measure of openness of the country according to Jonassen, 1950) is 377 suggesting quite open conditions. Unfortunately, pollen preservation was extremely poor in the old ground surface and underlying layers at Cashelkeelty II and so little information is retrievable. Keeping in mind the possibility of differential preservation, the pollen assemblage of the old ground surface suggests the presence of hazel scrub and heathland.

The interpretation of the pollen assemblages of the remaining podzol profiles can only be of the most general nature. It is impossible to define which spectra relate to the early stages of podzolization and some movement of pollen within the profile must have taken place during the stages of intense leaching and subsequent deposition of soluble metal oxides. Downwash of pollen from the overlying peat is considered to have been negligible as illustrated by the presence of many species in the upper mineral soil horizons, which are absent in the basal peat layers and *vice versa*. In fact, in any one soil profile, the pollen spectra are remarkably similar throughout, suggesting that no clear stratification exists. These spectra shall therefore be considered together in order to give a general impression of conditions prevailing *during* the period of podzolization.

At Cashelkeelty IV, considerable woodland appears to have existed in the locality while the podzol was developing. *Corylus* was the dominant species with *Ilex*, *Pinus*, *Betula* and a little *Quercus*. There is little evidence for heathland conditions. No radiocarbon date has been obtained for this profile because of the disturbed nature of the basal peat layer, but by comparison with the nearby Cashelkeelty I profile, the podzol development at Cashelkeelty IV was taking place towards the end of the Sub-Boreal (c 1200 bc) before the disappearance of *Pinus*. This means that within a distance of approximately 80 m, the timing of podzol development may have differed by as much as c 200 years and took place under different vegetational conditions. The only evidence of disturbance due to possible cultivation is where the basal peat is mixed with mineral soil, and since this must have occurred after peat growth began, and the podzol had fully developed, it cannot be put forward as a possible explanation for soil degradation. It could be that podzol development at Cashelkeelty II and III was delayed somewhat by reason of their location on an isolated elevation (1-2 m above surrounding land) with better drainage conditions. This implies a local topographical influence on the *timing* of the onset of podzol development.

Conditions prevailing during podzol development at Cullenagh I appear to have been grassland under cultivation and pastureland, with some heathland development and open scrubland dominated by hazel and birch. At Dromatouk heathland dominated with some open woodland. At Dromteewakeen similar conditions prevailed and in spite of the soil degradation, cereal cultivation was being carried out.

Maughanasilly I is the only site where an ancient tree cover can be demonstrated for the actual site examined. An extensive system of root channels was uncovered during excavation, mainly spreading horizontally through the upper few centimeters of the gleyed (G) horizon, and in a few instances reaching down into the top of the orange B_2 horizon. These channels had been given a uniform grey colour by the gleying process indicating their existence prior to the pre-peat waterlogging episode. Charcoal from the channels has been identified as Pinus sylvestris (Scots pine) and spreads of charcoal, also of Pinus sylvestris, from the surface of the G horizon have been dated to 3265 ± 55 bp (GrN-9281) or 1315 ± 55 bc. Charcoal embedded in the overlying basal peat gave an identical C^{14} date indicating that only a short period of time elapsed between the burning of the tree-cover and the onset of waterlogged conditions which gave rise to the growth of peat. The causal relationship between the two events however is not clear, but it does seem probable that the regional podzolization had in fact been developing under the cover of the pine trees.

It is of interest to note that even though pine charcoal in root channels was recovered from the top of the G horizon, no pine pollen was recorded for that level (Fig. 22). The concentration of pine pollen at the bottom of the G horizon and in the B_2 horizon suggests downward movement of the pollen, probably during leaching.

The above outline illustrates the complex nature of trying to reconstruct vegetational conditions at the onset of podzolization and during the course of its development. In the study area, woodland cover was reduced to open scrubland consisting of Corylus and Betula and extensive heathland tracts dominated by Ericaceae were in evidence when the podzolization process began. It is probable that this soil degradation was largely due to interference with the natural woodland cover. There is considerable evidence for cultivation throughout the podzolization period (on the basis of the presence of cereal pollen) and such disturbance of the soil surface coupled with factors such as the acidic parent material and oceanic climate, would have greatly encouraged the leaching process once started. It is however impossible to isolate one particular factor which was wholly responsible for the initial degradation of the humus layer. The timing of the podzolization process indicates that soil degradation had set in at different times in different localities, but generally speaking, would not have become a problem for cultivation until middle/late Bronze Age times (c 1000 bc).

5.3 Dating of peat initiation

With the exception of Cashelkeelty IV, the initiation of blanket peat growth at all sites was radiocarbon dated with the following results:

Table 1: Radiocarbon dates for initiation of blanket peat growth in SW Ireland.

	Number	Site	RCY bp	RCY bc/ad	Alti- tude	Material Dated
	GrN-9281	Maughanasilly I	3265 ± 55	1315 bc	213 m	Charcoal in Basal Peat
	GrN-9174	Dromatouk I	2020 ± 30	70 bc	107 m	Basal Peat
	GrN-9279	Cashelkeelty III	1060 ± 30	890 ad	110 m	Basal Peat
	GrN-9283	Dromteewakeen I	985 ± 45	965 ad	101 m	Basal Peat
x)	GrN-9282	Cullenagh I	975 ± 45	975 ad	244 m	Basal Peat
	GrN-9171	Cashelkeelty II	800 ± 30	1150 ad	101 m	Basal Peat

x) Sample treated rigourously due to possible root contamination.

At Cashelkeelty I (Fig. 16), the transition from soligenous woody peat to ombrogenous blanket peat formation is considered to have taken place at c 1100 bc when the Ericaceae and Cyperaceae (sedges) begin to expand and characteristic blanket peat species such as *Narthecium ossifragum* and *Myrica gale* appear.

During the past decade many radiocarbon dates have been obtained for the onset of blanket peat growth in various parts of Northern Ireland. These dates are summarised in Table 2 to facilitate comparison with those just obtained for the southwestern part of the country.

It is clear from these tables that the dates for the onset of blanket peat growth in Ireland are fairly evenly spread from the late third millennium bc right up to the twelfth century ad, with one possible gap between 85 bc and 890 ad within which at the moment, only one date falls (UB-367). It is tempting to correlate this gap with the drier, warmer climate which has already been suggested for the first millennium ad (see above, p. 4), but it may well be that when further sites are examined, this apparent anomaly will be removed.

Generally speaking, the dates from SW Ireland are later than those from the northern counties, a fact which may be due to topographical variations as has already been noted at a more local level, at Cashelkeelty (p. 57). It is also clear that altitude had little effect on the timing of blanket peat initiation. This point is well illustrated by the blanket peat deposits at Loughros Head and Slieve Gallion which are 30 m and 427 m above sea-level respectively but whose basal layers are both dated to c 2200 bc.

5.4 Factors Influencing Peat Initiation

The three main variables which are usually invoked to account for blanket peat formation are soil maturation, climate and human interference with the vegetation cover. The soil formation processes leading to peat growth in the study area have already been discussed in detail earlier in this chapter. At

Table 2: Radiocarbon dates for the initiation of blanket peat growth in N Ireland.

Number	Site	RCY bp	RCY bc/ad	Altitude	Reference
UB-2096	Loughros Head	4205 ± 80	2255 bc	30 m	Pearson, 1979
UB-274	Slieve Gallion	4165 ± 80	2215 bc	427 m	Smith, Pearson & Pilcher, 1971
UB-380	Lough Lark	3955 ± 75	2005 bc	329 m	idem
UB-158	Behy	3930 ± 105	1980 bc	150 m	Smith & Pilcher, 1973
UB-153	Behy	3890 ± 110	1940 bc	150 m	idem
UB-346	Beaghs sand quarry	3705 ± 65	1755 bc	278 m	Smith, Pearson & Pilcher, 1971
UB-376	Glens Bridge	3610 ± 75	1660 bc	232 m	idem
UB-364	Gruig Top	3335 ± 70	1385 bc	275 m	idem
UB-828	Slieve Croob	3325 ± 75	1375 bc	560 m	Pearson & Pilcher, 1975
UB-936	Butter Mt.	3270 ± 45	1320 bc	490 m	Pearson, 1979
UB-193	Pubble, ditch	3135 ± 90	1185 bc	183 m	Smith, Pearson & Pilcher, 1971
UB-337	Loughermore	2900 ± 70	950 bc	183 m	idem
UB-405	Crocknamoyle	2815 ± 90	865 bc	229 m	idem
UB-196	Pubble	2775 ± 75	825 bc	183 m	idem
UB-332	Altnahinch	2725 ± 85	775 bc	----	idem
UB-347	Beaghs Forest	2520 ± 70	570 bc	320 m	idem
UB-261	Beaghmore	2035 ± 70	85 bc	195 m	Smith, Pearson & Pilcher, 1970
UB-367	Breen Bog	1485 ± 65	465 ad	275 m	Smith, Pearson & Pilcher, 1971

some stage during the regional podzolization, a rise in water table occurred which resulted in the formation of an iron pan layer at that level with subsequent water-logging (and gleying) of the overlying soil, which in turn provided the conditions necessary for the invasion of the oligotrophic plants characteristic of the blanket bog vegetation. The crucial question therefore is, what caused the rise in water-table?

The role of climate as a controlling factor in peat initiation in SW Ireland is difficult to assess because of the lack of details regarding local conditions during prehistoric and early historic periods. If we accept Lamb's climatic reconstructions (see above p. 4), then peat formation in the study area began during periods of relatively dry and warm weather with little happening during the centuries centering on 600-400 bc for which a sharp climatic deterioration has been postulated! There is of course, the added problem of the diachronous nature of the basal layer of each tract of peat, which can result in a wide range of dates being obtained for a single tract, depending on how close the sample site is to the focal point at which peat formation started. This problem is well illustrated by the dates obtained for woody peat/blanket peat transition at Cashelkeelty I (c 1100 bc) and blanket peat initiation at Cashelkeelty II (1150 ad), sites which are but 100 m apart, on the same mountain ledge. It is probable that the focal point for blanket peat formation at Cashelkeelty is at the inner (southern) side of the ledge where the hillside rises steeply (and from which point the profile Cashelkeelty I was recovered) and that the peat gradually spread outwards to the northern edge of the ledge where the monuments are located and profiles Cashelkeelty II and III were taken. This would account for the 2000 year time-lapse between blanket peat formation at both locations. It should therefore be kept in mind that some of the dates quoted above for onset of peat formation may refer to points nearer the margins of the peat tracts rather than their focal points and so may lead to misleading conclusions.

The role of man in initiating or hastening the formation of peat has been discussed by Mitchell (1972, 1976), Moore (1973, 1974 and 1975) and Goddard (1971). The recognition of former farmland under blanket peat deposits in Northern Ireland, led Mitchell to speculate on the possibility of a causal relationship between prehistoric farming activities and the onset of blanket peat growth. He suggested that disturbance of the upper layers of the mineral soil by ploughing or spade cultivation may have made the soil more susceptible to waterlogging with subsequent iron pan formation and peat growth. Alternatively, he suggested that deteriorating climatic conditions may have been the primary factor, with man's disturbance of the soil cover merely accelerating the process.

These ideas were later taken up by Moore who examined specific sites in Wales and England in an effort to throw some light on the problem. By using pollen analysis on a shallow peat deposit in Merioneth, Wales, Moore illustrates that during the period of peat initiation several interference phases can be recognized during which changes in the local pollen, both in diversity and specific composition, took place. On the basis of these changes, it is concluded that "human activity, such as scrub or canopy clearance, firing or stock grazing, could have accelerated the degeneration of a damp woodland ecosystem and assisted the spread of blanket bog" (Moore, 1973, 352). Similarly, in the case of peat profiles from Exmoor, Moore (1974, 441) concluded that it was man, or his domesticated animals which provided the final stress on an eco-system already under climatic strain and so brought about the onset of peat growth. He also emphasizes the point that human clearance of woodland can modify hydrological regimes substantially and so prehistoric man was theoretically capable of influencing catchment hydrology "in ways which would create effects of magnitude similar to those derived from major climatic changes" (Moore, 1975, 269).

Moore's conclusions linking human activity with the onset of peat growth, are based on two types of evidence; the dating of basal peat layers at about the time of arrival of farming cultures and indications of considerable human activity in the basal deposits. While it is true that several basal peats have been dated to early Neolithic times there are just as many again whose dates range through the following millennia, even when Neolithic or Bronze Age activity has been demonstrated for those sites. The extreme case of Cashelkeelty II may be quoted where human activity on the site was dated to 970 bc but the overlying basal peat gave a date of 1150 ad. Such problems concerning the dating of basal peat layers have already been discussed (p. 61). Also the uncertain time-lapse between the rising of the water table on a site and subsequent peat formation means that the activities registering in the basal peat layers may in fact have taken place some considerable time after the initial unbalancing of the eco-system which set in motion the processes leading to peat formation. It is therefore suggested here that the pollen evidence from the underlying mineral soils should be more informative than basal peat layers, for the reconstruction of events leading up to peat growth.

It has already been shown (p. 58) that woodland cover was greatly reduced at each site in the study area by the time that podzolization and peat growth took place. This removal of woodland cover was almost certainly due to man's activities but there is no definite proof to link this event with the rise in water-table which eventually caused the onset of peat growth. At Dromteewakeen I, and Cullenagh I there is considerable evidence for cultivation activities in the upper mineral soil horizons and it may be that here we have a case of man's disturbance of the soil being the primary causal factor for peat growth. This is merely a suggestion however, in the absence of a date for the cultivation episodes. At Cashelkeelty II and III the peat growth is obviously the result of marginal spreading of a peat mass with its focus in the vicinity of Cashelkeelty I where the natural progression from soligenous woody peat to ombrogenous blanket peat took place under conditons of high precipitation/evaporation. At Maughanasilly I the burning off of the pine tree vegetation on the site itself at c 1315 bc may well have caused drastic hydrological changes which facilitated the onset of peat formation — certainly there was little time-lag between these two events since some of the pine charcoal was incorporated in the basal peat layer. Finally, at Dromatouk I there is evidence of burning pine at 1380 bc but peat formation did not begin on the site until 70 bc. It is not clear whether this pine charcoal is the result of firing a stand of pine trees _in situ_ or whether it simply represents wood brought onto the site from elsewhere and so no conclusion may be drawn regarding man's possible role in assisting the initiation of this particular peat deposit.

It should be clear from the arguments just outlined, that no one factor may be put forward as the critical one responsible for the origin of blanket peat growth. As with the podzolization process, a complex network of factors is involved and any one of these factors (e.g. climatic deterioration or removal of woodland cover), may have put the final stress on an already unstable eco-system of a site. The role of climatic deterioration cannot be ignored but it is also suggested that at certain sites, man's disturbance of the tree cover or soil surface was a contributing factor to the onset of blanket peat formation. Similar conclusions were reached by Goddard (1971) for the initiation of blanket peat growth in Northern Ireland and in this area, the effects of human interference were not considered very drastic until Bronze Age times.

5.5 <u>Summary</u>

In this chapter two main problems are discussed: the processes leading to the formation of the gleyed podzol profiles at each site and the dating and possible influencing factors for the intiation of blanket peat growth in the study area.

Podzolization was beginning in the study area by the early second millennium bc but in some localities the process was initiated much later. At most sites examined, the podzols were forming under conditions of open heathland with some *Corylus* and *Betula* scrub. In spite of obvious soil degration, cultivation was still being carried out at several sites. No one factor could be invoked to account for the initiation of podzolization in the study area but once started, factors such as acidic parent material, oceanic climate and disturbance of the soil surface would have greatly encouraged the leaching process.

Dates obtained for the initiation of blanket peat growth in the study area range from c 1315 bc to c 1150 ad. It was concluded that the interaction of several different factors was responsible for the onset of peat growth. It is suggested that at several of the sites examined, man's activities in the form of burning off the woodland cover and/or disturbing the upper soil horizons may have been responsible for causing or simply accelerating the iniation of peat growth. The role of climatic deterioration in causing the rise in water table which ultimately led to peat formation, must also be considered. Altitude apparently had little effect on the timing of the onset of peat growth and it is likely that local topographical factors may account for some of the variations.

To prove a causal relationship between human activities and the onset of peat growth is extremely difficult and whereas this study has suggested disturbance/cultivation of the soil as a possible <u>contributing</u> factor, a complex interplay of events and local conditions must be seen as responsible.

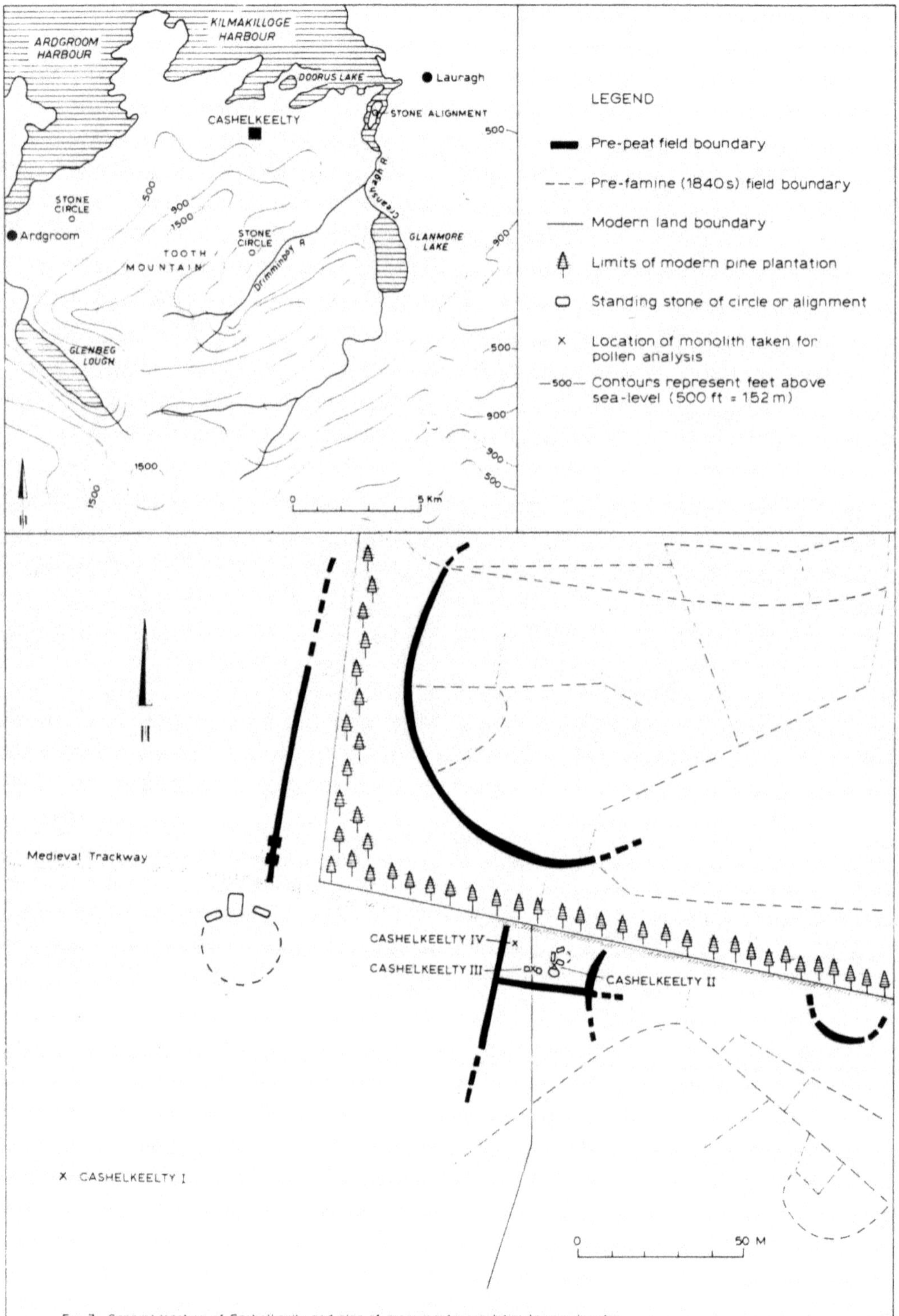

Fig 7 General location of Cashelkeelty and plan of monuments in relation to sample sites.

CHAPTER 6

THE EXCAVATED SITES

6.1 Introduction

It is proposed to outline here the general results of excavations which were carried out as part of this study. Since these excavations are to be published elsewhere in detail (Barber, forthcoming a; Lynch, in press), only those aspects which are of relevance to this particular study shall be considered. The sites chosen for excavation were (i) Cashelkeelty, Co. Kerry where the close proximity of stone circles, alignment and field fences offered the possibility of determining their cultural and temporal relationships, and (ii) Maughanasilly, Co. Cork which is one of the best preserved examples of a stone alignment. Both excavations were carried out in the period May-August 1977.

6.2 Cashelkeelty Stone Circles and Alignment

Cashelkeelty is located at an altitude of 110 m OD on the northern side of the Beara peninsula, approximately 20 km west of Kenmare town and overlooking Kilmakilloge harbour (Fig. 7). The site before excavation consisted of an alignment of three large standing stones, c 2 m to the north of which were the recumbent and two standing stones of a stone circle. Both of these monuments are sited on a localised elevation which is 3-4 m higher than the surrounding area. In the immediate vicinity of these monuments, traces of early field fences could be discerned, partly overgrown by peat. About 80 m to the west of the alignment and circle, the remains of another monument in the form of two large standing stones and one fallen, could be seen (Fig. 7). This complex of monuments is sited on a level ledge (maximum dimensions c 500 m long x c 300 m wide) on the peat-covered northern slopes of Tooth mountain which descends steeply to the narrow coastal plain (100-200 m wide) of the Kenmare estuary. Just to the north of the monuments described, remains of the medieval roadway from Kenmare to Castletownbere may still be seen.

6.2.1 The Excavation. The area excavated (in total c 1000 m^2) was centered on the features visible above the blanket peat cover and was so designed as to give maximum information about the following points:

(a) the original form of the monuments;
(b) the date of construction of the monuments;
(c) the possible cultural relationships between the different momuments;
(d) the location and form of associated settlements;
(e) the general sequence of events during the prehistoric period on this ledge.

6.2.2 Results. Reconstruction of the sequence of events on the site was greatly hampered by the gleying process (already described in chapter 5) which had reduced the upper levels of the mineral soil to a uniform whitish/grey colour, thereby masking details of stratification. The following sequence which is suggested for the features uncovered is therefore based not only on archaeological stratigraphy, but also on the stage of soil formation reached (i.e. the degree of podzolization or gleying) at the time of construction.

Phase 1. On the basis of pollen analysis of old ground surfaces (described in detail in chapter 7), the earliest activity on the Cashelkeelty ledge was represented by a period (or periods) of cereal cultivation and possible pastoralism which took place before c 1000 bc.

Phase 2. The stone alignment was constructed c 1000 bc after the onset of podzolization but before gleying took place. This date suggested for the construction of the alignment is based on stratigraphic evidence which indicated a sufficient time lapse between the construction of the alignment and that of the circle in the period 970-715 bc to allow the regeneration of a thin humus layer over disturbed soil. The alignment originally consisted of four standing stones with the socket of the fourth stone lying to the east of the three now standing. This fourth stone was removed at some stage before the growth of peat on the site (c 890 ad).

Approximately 1 m south of the alignment, a large pit (1.5 m x 0.5 m x 0.6 m deep) covered by flat slabs was revealed. No finds were recovered from this pit which had gradually filled up with loose soil washed in from above, but its shape and dimensions are suggestive of a grave. The construction of the pit took place before the gleying and it is possible that both alignment and pit belong to the same activity phase.

Phase 3. The recumbent stone circle was constructed 1.40 m north of the stone alignment sometime in the period 970-715 bc. This circle originally consisted of a recumbent with four standing stones and had an internal diameter of 1.5 m. In the centre of the circle, a pit containing a cremated burial was covered by a single slab. This burial was in a primary position within the circle and on analysis, was found to be probably that of an individual, aged between 25-30 years.

Two radiocarbon dates were obtained which relate to the construction period of the circle: (a) 2920 ± 60 bp (GrN-9173) or 970 bc, from charcoal embedded in the old ground surface which was sealed by a layer of stony soil used to level the sloping surface before construction of the circle and (b) 2665 ± 50 bp (GrN-9172) or 715 bc from charcoal found in the stony soil layer itself. This latter charcoal could however be a later intrusion, with all signs of the disturbance having disappeared because of the gleying. Consequently, the construction of the stone circle may be assigned to the period 970-715 bc but without more precise dating.

The distribution of peat covered field fences on the Cashelkeelty ledge is shown in Fig. 7 which indicates clearly that no field system as such may be postulated on the basis of surface indications — in fact not even one completely enclosed field may be traced. It is therefore impossible to say whether or not these separate fence segments belong to the same system or whether they represent remains of several different episodes of land enclosure. Even on excavation it can be extremely difficult to establish the chronological relationships between abutting fences or even separate segments of a continuous line (see Barber, in preparation b).

However, a 14 m length of field fence was excavated just south of the Cashelkeelty alignment in an effort to retrieve some dating evidence. The foundation course survived in patches only, with many boulders and slabs representing tumble from the wall, lying on the downhill side. No direct evidence for dating this stretch of field fence was recovered but its construction must have predated both the gleying process and peat formation. The chronological relationship between the field fences and other monuments at Cashelkeelty therefore cannot be accurately defined but it is tentatively suggested that at least some of the segments belong to the cultivation phase(s) which predate the construction of the alignment and circle (see chapter 7).

Extensive area excavation in the vicinity of the two standing stones at the western end of the ledge, revealed that they originally formed part of a multiple-stone circle (internal diameter c 17 m) containing 11 or 13 stones. This circle had been almost completely destroyed by the time peat began to form on the site and many subsequent disturbances in the region of the remaining two standing stones may be attributed to the period of use of the nearby medieval trackway. Apart from the sockets of the standing stones, no other features were revealed.

No traces of a settlement, either in the form of structures or features, were recovered during the entire excavation.

A total of seven artefacts were recovered — six flint implements and part of what is possibly an ard point, made of sandstone (Fig. 8). It was originally thought that the latter artefact could have been the butt of a stone axe but the clearly faceted edges and wear marks are more suggestive of an ard point, in particular of the type found on the Northern Isles of Scotland as described by Rees (1979). The leafshaped arrowhead (Fig. 8, b), scraper (Fig. 8, e) and possible part of a hollow scraper (Fig. 8, c) were all found in the top of the gleyed horizon in the vicinity of the stone alignment. The barbed and tanged arrowhead (Fig. 8, a) and side scraper (Fig. 8, f) were also found near the alignment but at the bottom of the gleyed horizon. The flint point (Fig. 8, d) and fragment of ard point were found resting on top of the stony layer deposited during construction of the recumbent stone circle. The only artefact recovered from the multiple-stone circle was a flint scraper (Fig. 8, g) found in the top of the gleyed horizon. In addition to these artefacts, six small unworked flint flakes were found.

Again, the changes brought about in the soil profile by the gleying process have made it virtually impossible to determine the association of the finds. Only the ard point fragment and flint point (Fig. 8, d) may be tentatively assigned to the construction period of the five-stone circle but the possibility of a later date cannot be precluded. On the basis of old ground surface levels and the uppermost level of packing stones in sockets it is most likely that the activity level of the alignment and circle builders lies within the upper few cm

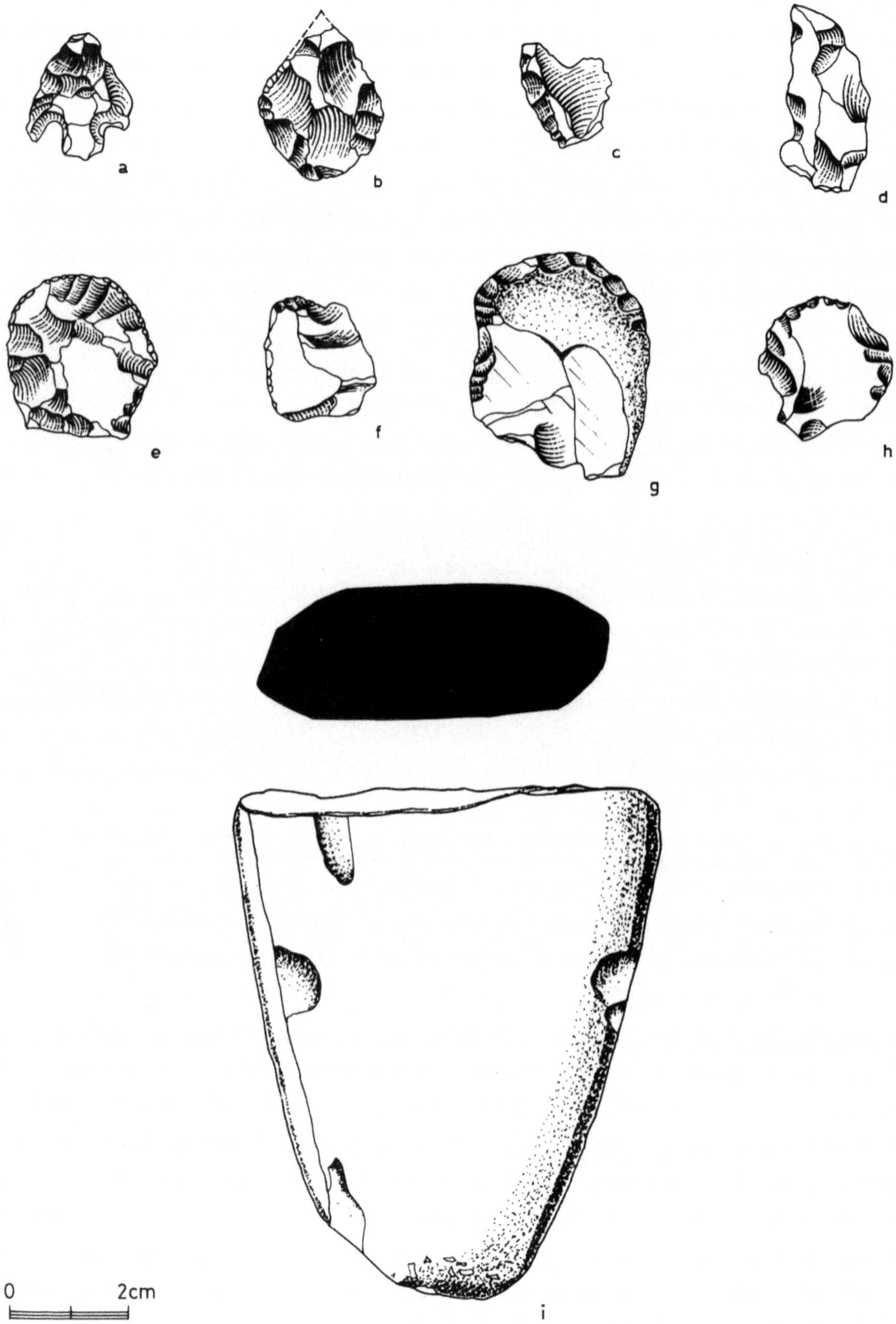

Fig.8 Finds from Cashelkeelty (a, b, c, d, e, f, g, i) and Maughanasilly (h).

of the gleyed layer. It is therefore possible that the barbed and tanged arrowhead and scraper (Fig. 8, a, f), both found at the bottom of the gleyed horizon, belong to an earlier phase of activity.

The range of artefacts as a whole is relatively undiagnostic, and could belong to any period during the Neolithic or Bronze Age. The stone ard points of Britain may be found in early Bronze Age or later contexts and while barbed and tanged arrowheads are considered characteristic of Beaker or early Bronze Age sites, they may also occur in middle or late Bronze Age contexts (Herity & Eogan, 1977, 148). On the other hand, leaf-shaped arrowheads are usually found in Neolithic contexts.

6.3 Maughanasilly Stone Alignment

The Maughanasilly alignment is located at an altitude of 213 m OD in the valley of a tributary of the Coomhola river, c 17 kilometres northeast of Bantry town (Fig. 9). The alignment of five standing stones is sited on the crest of a small boulder-clay hillock (now covered with blanket peat) with Knockbreteen mountain (243 m OD) rising steeply to the south. Lough Atóoreen lies c 300 m to the southwest while Knockboy and Conigar mountains (c 230 m OD) border the valley to the northwest (Fig. 9). These mountains are part of the Shehy range and local bedrock is Old Red Sandstone. Extensive tracts of blanket peat are present in the vicinity of the stone alignment.

6.3.1 The Excavation. With the exception of a recently disturbed area at its western end, the entire hillock crest was excavated (c 170 m^2). The excavation was carried out with three main objectives in mind:

(i) To detect any signs of occupation in the form of structures or artefacts which could be associated with the alignment;

(ii) To recover dating evidence for the construction of the alignment;

(iii) To obtain a series of samples suitable for pollen and seed analysis and thereby place the alignment in its ecological context (see chapter 7).

6.3.2 Results. Similar difficulties to those just outlined for Cashelkeelty (p. 66) were encountered when trying to determine relationships and association on the basis of stratigraphy greatly modified by leaching and gleying (for details of the soil profile see p. 52). Consequently even though it may be assumed that all five standing stones were erected at the same time, not even this can be proved conclusively. However, wherever packing stones in sockets were present, the topmost stone was usually just flush with the top of the gleyed horizon which suggests that the sockets were dug from the upper levels — if not the very top — of this gleyed layer, which may therefore be regarded as the activity level associated with the construction of the alignment.

There is no evidence to suggest that the alignment ever consisted of more than the five standing stones remaining today. Four small pits and a possible post-hole were revealed in the vicinity of the alignment and may well have served some function during its construction. Further evidence of prehistoric activity on the hillock was scarce. Several charcoal spreads occurred in the upper levels of the gleyed horizon which may be partly explained as the result

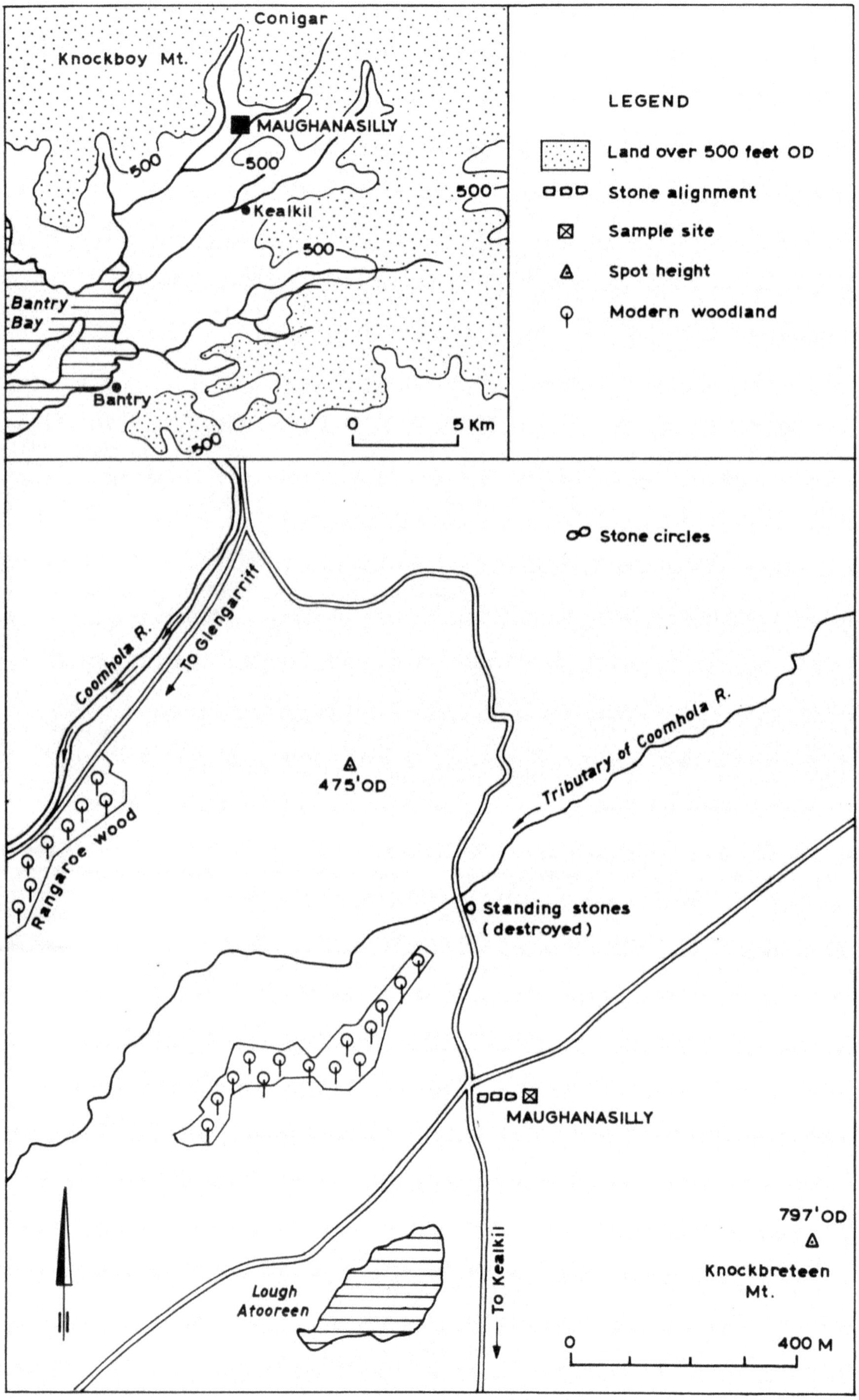

Fig.9 Location of Maughanasilly alignment and sample site.

of burning trees or shrubs since several pre-gley root channels were found, which contained large amounts of charcoal. The charcoal from the root channels was identified as that of Pinus sylvestris and gave a C^{14} date of 3265 ± 55 bp (GrN-9280) or 1315 bc while some Quercus and possible Alnus was represented together with Pinus sylvestris in the remaining spreads.

Charcoal embedded in the basal peat overlying the site gave an identical C^{14} date to that recovered from the root channels: 3265 ± 55 bp (GrN-9281) or 1315 bc, which suggests that only a short period of time elapsed between the burning of the tree-cover and invasion of the hillock surface by blanket peat which incorporated some of the remaining charcoal in its basal layer. It was clear during excavation that construction of the alignment took place before the gley and peat formation and so it is likely that the burning episode took place during the alignment construction (perhaps as part of site clearance prior to erection of the standing stones) with peat growth starting soon afterwards. The date of 1315 ± 55 bc may therefore be considered as more than likely relating to the construction of the alignment.

The settlement area of the alignment builders was not located. Only one find, a thumb-shaped flint scraper (Fig. 8, h) was recovered throughout the entire excavation. Such scrapers are commonly found in Bronze Age contexts throughout Ireland.

6.4 Discussion

The excavations at Cashelkeelty and Maughanasilly were undertaken primarily to clear up questions regarding the dating and cultural/ecological context of the stone circle/alignment complex in SW Ireland. As already stated (p. 19), the few stone circles previously excavated had produced a small number of nondescript finds and the rather dubious C^{14} date of 600 ± 120 ad, while the stone alignments had remained unexcavated. Even though the present excavations did not succeed in answering all the questions posed, they have at least, placed the stone circle/alignment complex in a firm temporal and environmental context and have suggested a direction in which future research, might provide a fuller understanding of the momuments and period in question. The information provided by the excavations is discussed briefly below while the wider implications of the results are considered together with the environmental evidence in chapter 8.

The number of stones, revealed by excavation, in the Cashelkeelty and Maughanasilly alignments, is in good agreement with the SW Ireland group in general (maximum of 6 stones). The morphology of the Cashelkeelty circles (one five-stone circle and one 11- or 13-stone circle) also agrees with the grouping outlined by Barber (1972) and Ó Nualláin (1975) which has been described above (p. 19).

The purpose of the stone circles and alignments remains obscure. The cremation burial found at Cashelkeelty is the first evidence of such burials associated with a five-stone circle (all previous instances of burials occurred in multiple-stone circles) and it is possible that the larger multiple-stone circle also contained a burial all traces of which would have disappeared during subsequent interference with the momument. The grave-like feature probably associated with the Cashelkeelty alignment may once have held an inhumation but no evidence was recovered at Maughanasilly to suggest either inhumation or cremation burials in the vicinity of the alignment.

The presence of burials within stone circles certainly indicates that these monuments had some connection with funerary activities but it does not explain their purpose or function. In the case of the alignments, the picture is even more hazy and one can speculate endlessly about ritual or symbolic significance, territorial markers or even routeway markers. It is doubtful whether future excavation of these monuments alone will ever provide the answers, since by its very nature, evidence of ritual or symbolism is virtually irretrievable.

One of the major aims of these excavations was to try to locate the habitation sites associated with the monuments and thereby gain some information regarding the cultural and economic background of these communities. Failure to locate these habitation sites during excavation suggests that they may be sited outside the immediate vicinity of the alignments and circles. Extensive field-walking, and in the case of Cashelkeelty, full aerial photographic coverage of the area, did not succeed in locating possible settlement sites, all of which suggests that the settlements were of a flimsy nature, the remains of which now lie hidden beneath the rolling bogland. An extensive programme of systematic boring and chemical analysis of the pre-peat mineral soil could perhaps offer a solution to the problem.

One can however, speculate as to the possible siting of the Cashelkeelty and Maughanasilly settlements. The investment of labour represented by the Cashelkeelty monuments suggests that a fairly large group, probably representing several families must have been involved in their construction. Even though the mountain ledge on which the monuments are sited was certainly under cultivation prior to and perhaps contemporary with their construction (p. 87), the area represented (max. 150,000 m^2) would hardly have supported a large community. It is therefore possible that these monuments were built by community effort on farmland whose soils were already degenerating (as evidenced by developing podzolization) and that the major areas of settlement were elsewhere. Two areas in the immediate locality would have provided suitable habitation sites. The first is the coastal plain and in particular the Doorus peninsula (Fig. 7), which may have been more extensive during the period in question due to lower sea-levels (see p. 3). Such locations would have the obvious advantages of proximity to harbourage, fishing, and fresh water supplies (i. e. Doorus Lake) and the availability of a certain extent of level land suitable for cultivation with the nearby hillsides available for pasturage and some additional cultivation. The stone alignment on Knocknacappul island c 1.5 km to the east (Fig. 7) supports this idea of activity on the coastal plain. The other probable area of settlement is the valley of the Creanshagh and Drimminboy rivers which extends for c 2.5 km north-south, at a distance of c 1.5 km to the east of Cashelkeelty (Fig. 7). This narrow valley in an otherwise mountainous terrain must have been exploited in Bronze Age times as evidenced by the Shronebirrane stone circle near the headwaters of the Drimminboy (Fig. 7). It is therefore suggested here that the Cashelkeelty ledge represents an area which was marginal to the main focus of settlement and once its cultivation potential began to decline it was chosen as a site for funerary activities. Instances of burial/ritual monuments being built on what was previously agricultural land, are common throughout prehistory, and this subject shall be returned to later (p. 94) when the environmental evidence provided by the old ground surfaces is considered.

The location of the Maughnasilly settlement is just as elusive as that of Cashelkeelty. Extensive fieldwalking was undertaken in the region of the alignment in an effort to locate traces of habitation but again, the blanket peat cover has effectively concealed all such traces. Considering the local topography however, it is probable that some settlement took place in the rather narrow valley of the Coomhola tributary which runs northeast/southwest and which contains in addition to the Maughanasilly alignment, standing stones (now destroyed) and two stone circles, c 400 m and c 1 km respectively, to the north (Fig. 9).

This valley would have provided a relatively fertile pocket of land in the middle of what is otherwise barren mountainous terrain. Lough Atooreen at the southern edge of the valley would have provided fishing opportunities and a convenient freshwater supply and it is not improbable that the lake-shores themselves were the focus of settlement. Access to the sea (c 10 km distant at Bantry Bay) would not have been difficult via the Coomhola river.

Perhaps the most significant outcome of these excavations is the dating evidence recovered for the construction of the stone circles and alignments. The building of the Cashelkeelty five-stone circle is bracketed by the dates 970-715 bc which places it in the late Bronze Age period. The dating of the Cashelkeelty and Maughanasilly alignments is less precisely defined, but the period c 1300-c 1000 bc is probable. It is of interest to note here that a thick charcoal layer from the upper gleyed horizon near the Dromatouk alignment (p. 100) has given a C^{14} date of 3330 ± 50 bp (GrN-9346) or 1380 bc. By analogy with the excavated sites, it may be argued that the stratigraphic level of this charcoal layer suggests association with the construction period of the nearby alignment, and the date of 1380 bc is in general agreement with those suggested for Cashelkeelty and Maughanasilly. The monoliths taken from Cullenagh and Dromteewakeen did not provide any dating evidence which could be associated with the alignments in the vicinity except that the basal peat dates of 975 ad and 965 ad respectively, post-date their construction.

If one assumes, as is most likely the case, that the alignments and circles are part of the same cultural complex, then the results presented here indicate a date-range of c 1400-c 700 bc for their construction and use. This date-range is remarkably similar to that of 1535-775 bc obtained by Pilcher (1969, 81) for the construction of the circles and alignments at Beaghmore, Co. Tyrone. It is perhaps of interest to note here also, that on the basis of astronomical orientations, a date range of c 1800-600 bc was suggested by Barber (1973, 35) for the stone circles of SW Ireland and the period 2000-1000 bc was put forward by Ní Loingsigh (1976, 134) for the building of stone alignments. When considering the cultural context of these monuments then, we are dealing with the later stages of the Bronze Age (for details see p. 18). Urn and food vessel burial traditions still survived, the metal industry was rapidly developing both typologically and technologically, greater emphasis was being put on ornaments and hoards containing large amounts of metal were being deposited. In the study area the later Bronze Age is represented almost exclusively by the hoards of bronzes and gold described above (p. 23). The copper mines at Mount Gabriel were being worked at this time and it is reasonable to suggest that mining was also taking place at many of the other copper-bearing deposits in the region, although direct evidence for such activities has not yet been obtained. When considering together the distribution of the stone circle/alignment sites in relation to the copper deposits (Fig. 3) and the presence of many late Bronze Age metal hoards in the study area, it is tempting to suggest that the builders of the monuments were connected in some way with the flourishing metal industry, whether at the mining, manufacturing or distribution stage. However, if such a connection did exist, one would expect some few metal pieces to have been found during excavation, perhaps associated with a burial. It may well be that metal was still such a precious commodity that it was ke in circulation for as long as possible, but on the other hand, the evidence of the hoards suggests quite the opposite if one accepts O'Shea's (1980) theory concerning hoard deposition as an inflationary device. For the moment therefore, there simply has not been enough raw data collected from these sites with which to test any hypotheses concerning their possible relationship to the Bronze Age metal industry.

6.5 Summary

This chapter deals in outline with the results of two excavations carried out as part of this study — the stone circles, alignment and ancient field fences at Cashelkeelty, Co. Kerry and the stone alignment at Maughanasiily, Co. Cork. The following conclusions were derived:

(i) The stone circles, and possibly also the alignments are closely connected with funerary activities but their precise function remains unknown.

(ii) These monuments were built some distance from the related settlements and sometimes on degraded farmland.

(iii) The construction period of the circles and alignments may be bracketed by the dates c 1400-c 700 bc which places them in the later part of the Bronze Age.

CHAPTER 7

THE VEGETATIONAL HISTORY OF SW IRELAND FROM c 4000 bc TO PRESENT

7.1 Introduction

A major part of this study is concerned with an attempted reconstruction of the changing vegetation patterns in SW Ireland since c 4000 bc, with particular emphasis placed on the evidence for man's interference with this vegetation cover during prehistoric times. Five sites were investigated (Fig. 4) and eight pollen and one macrofossil diagrams were prepared. The results of these analyses are presented here and the reader is referred to chapter 4 where the methods employed, are outlined.

The data are presented in the form of pollen diagrams (Figs. 16, 18-23), tables 8-14 and a macrofossil diagram (Fig. 17). The pollen diagram from the long profile, Cashelkeelty I is illustrated in full (Fig. 16) but for the shorter profiles only those taxa of relevance for the discussion are included in the diagrams and the remainder are presented in tables 8-14.

The code used in the pollen diagrams to illustrate the degree of humification of the peat is based on observations made in the field and when sub-sampling later in the laboratory. Generally speaking, light brown fibrous peat with many visible plant remains was classified as being of 'low humification', dark brown more compact peat was placed in the 'moderately humified' category and 'highly humified' peat implies a black greasy compact deposit with no visible plant remains.

As already mentioned (p. 46), local pollen assemblage zones have been used to sub-divide the diagrams for discussion purposes. The definition adopted for a local pollen assemblage zone is that used by West (1970, 1180), i.e. "a body of strata characterized by a certain assemblage of fossils without regard to their ranges." These pollen assemblage zones therefore reflect a vegetation unit or collection of units of local significance and are not related to chronostratigraphy. Regional assemblage zones for the study area have not been established since only one site, Cashelkeelty I, provided a sufficiently long profile to be radiocarbon dated at critical levels and in the absence of other such profiles it was considered premature to accept this as a type site. References shall be made to the earlier zonation schemes of Blytt (1876) and Sernander (1908), Jessen (1949) and Mitchell (1956) where relevant, and a correlation of all schemes is presented in Fig. 5.

Throughout the analyses every effort was made to differentiate between the pollen of *Corylus* and *Myrica gale*, since the latter species is widespread in the blanket peat areas of W Cork and Kerry and its possible high values could therefore greatly confuse the interpretation of the diagnostic *Corylus* curve. This problem of differentiation has for long been recognized by palynologists (Jessen 1949, 103; Pilcher 1968, 87; Simmons & Cundill 1974, 161) but little headway has been made in reaching a solution, short of using scanning electron microscopy. For this study, the identification criteria used are those established by palynologists with experience in dealing with these species, in particular Prof. dr W van Zeist, Biologisch-Archaeologisch Instituut, State University, Groningen (personal communication) and Dr. M Telford, Botany School, Trinity College, Dublin (personal communication). In general it was found that many of the diagnostic characteristics of Myrica gale grains (e.g. the 'pouting' pore) were more easily recognized in samples mounted in glycerol jelly than those in silicone oil (possibly as a result of the slight swelling of grains when mounted in the glycerol jelly). Only those grains which could confidently be assigned to the species Myrica gale are included in the Myrica curve while the doubtful cases remain in the Corylus curve. The frequency of the latter was not considered high enough however, to produce a significant bias.

7.2 Cashelkeelty, Co. Kerry

Four monoliths for pollen analysis were recovered from the immediate vicinity of the stone circles, alignment and field fence complex at Cashelkeelty, Co. Kerry (Fig. 7).

7.2.1 CASHELKEELTY I (Figs. 16, 17)

7.2.1.1 Site Description. Townland: Cashelkeelty Altitude: 101 m OD
County: Kerry Nat. Grid. Ref.: VO75 052

This monolith was recovered from a 3 m deep peat deposit on the southern side of the mountain ledge already described (see above, p. 65). This peat deposit had formed in a localised hollow in the bedrock at a point where Tooth mountain begins to rise steeply (Fig. 7) and was located by probing. Elsewhere on the mountain ledge, the blanket peat cover averages 30-40 cms in depth. The location of this profile is an extremely sheltered one since it lies on the leeward site of Tooth mountain and of the Caha mountain range in general and is thus protected from the prevailing southwesterly winds. The Kenmare River estuary which lies immediately north of Cashelkeelty is bounded on the north and east by the extensive mountain ranges of the Macgillycuddy's Reeks, the Derrynasagart mountains and the Shehy Range, while c 10 km due west of Cashelkeelty is the open Atlantic.

The pollen record from Cashelkeelty is therefore not likely to be greatly influenced by long distance transport by wind and should reflect both the local and extra-local vegetation patterns within the limits set by variations in the pollen production and dispersal of different species.

The present-day vegetation on the mountain ledge consists of typical blanket peat species such as Molinia caerula, Scirpus caespitosus, Narthecium ossifragum and Potentilla erecta (tormentil). Pteridium aquilinum is widespread at the eastern end of the ledge particularly in the abandoned pastures. Along the eastern approaches to the site, mainly bordering the medieval trackway, are scattered occurrences of Ilex aquifolium, Sorbus aucuparia (rowan), Crataegus monogyna (hawthorn) and Ulex galii (autumn gorse).

Immediately to the north of this trackway, covering the lower slopes of Tooth mountain and extending onto the Doorus peninsula (Fig. 7) is the plantation (mainly Pinus sylvestris) of the Landsdowne Estate dating to the early 1900s.

The Cashelkeelty hillside is today, given over entirely to sheep grazing on an extensive scale.

7.2.1.2 Sampling and Stratigraphy. The upper 1.5 m of the peat monolith was collected in February 1977 and the lower 1.5 m was recovered the following July when the extra manpower needed to extend the pit was available during excavation (see above p. 49 for details of method of collection). These monoliths overlapped by 50 cm, a fact which was noted in the field and subsequently recognized and corrected in the pollen diagram.

The stratigraphy at this point showed 3 m of peat resting directly on bedrock. Examination of the peat profile both in the field and later in the laboratory did not succeed in differentiating between the lower wood peat and the overlying blanket peat deposit which were subsequently recognized on the basis of the pollen analysis. The term 'wood peat' could perhaps be considered a misnomer since very few wood remains were in fact recognized in the peat, but as shall be seen later (p. 79) Salix carr (willow carr) must have been well developed in this swampy hollow before decreasing minerotrophic influences brought about a transition to the more oligotrophic blanket peat vegetation. The absence of Salix wood from this lower deposit may perhaps be explained by its rapid rate of decay (Prof. J Moore, personal communication).

1-7	cm	Fibrous light brown blanket peat with much root penetration.
8-28	cm	Dark brown, highly humified blanket peat.
29-33	cm	Gradual decrease with depth in humification of the blanket peat. Some charcoal fragments at 27.5 cm.
34-122	cm	Blanket peat of low humification, including some small wood fragments and possible Molinia remains.
123-136	cm	Highly humified blanket peat.
137-143	cm	Blanket peat of intermediate humification and a little wood.
144-148	cm	Strip of highly humified dark brown blanket peat, again with a few small wood fragments.
149-158	cm	Blanket peat of low humification. Two distinct layers of small branches at 154 and 158 cm which could possibly represent part of a wooden trackway. Unfortunately, due to the dangerous conditions produced by rising water level and slumping of the peat faces in the pit, this possible feature could not be investigated further.
159-c 182	cm	Blanket peat of intermediate humification.
c 183-c 190	cm	Transition to wood (willow carr) peat with a little charcoal at 190 cm.
c 191-200	cm	Peat of intermediate humification with wood and charcoal fragment s at 199 cm.
201-240	cm	Dark brown, slightly humified peat with some twigs and several charcoal streaks at 210 cm and between 235-240 cm.

241-250	cm	Moderately humified peat with some wood and charcoal streaks at 242 cm.
251-290	cm	Black, highly humified peat with a little wood at 270 cm.
291-300	cm	Very moist, black peat with a little wood at 292 cm.
> 300	cm	Bedrock.

7.2.1.3 Presentation of Data. The pollen values were calculated as percentages of total pollen excluding *Salix* (because of its extremely high values in the lower wood peat) and the spores of ferns and other lower plants. The latter were calculated as percentages of that sum. The pollen samples were analysed at 5 cm intervals, except in the region of the possible 'trackway' where the interval was reduced to 2 cm. Samples (c 30 ml) were taken from the monolith at 10 cm intervals and analysed for macrofossil remains such as seeds, leaves, wood etc. which could give added information about the local vegetation. The results are presented in Fig. 17 and so little information was gained in proportion to the amount of time spent on analysis that it was decided not to sample at closer intervals. The identified remains consist almost entirely of seeds. Some leaves of *Sphagnum* and a few carbonised flowers of the Ericaceae family were also identified. The majority of wood and charcoal fragments were too small for identification.

7.2.1.4 Radiocarbon Dating. A total of eight radiocarbon dates were obtained for different levels of the Cashelkeelty I profile. All samples were processed by Dr G W Pearson and his staff at the Palaeoecology Laboratory, Queen's University, Belfast. The majority of the samples consisted of trimmed 2 cm thick slices removed from the appropriate levels of the metal boxes but in the case of UB 2368, UB 2413 and UB 2414, a 4 cm slice was necessary to provide sufficient carbon for dating. In these latter instances where additional material was required, a 1 cm slice was submitted from immediately above and below the original 2 cm slice. In all cases the whole peat (A fraction *sensu* Dresser,1970, 35) was dated, following acid pretreatment. Comparison between these dates and those obtained from the Groningen Laboratory for the remaining sites, should be valid, since similar pretreatment was carried out and both laboratories have recently calibrated the new Oxalic standard and agreed to within c 15 years (Pearson, personal communication). The following is a list of Cashelkeelty I radiocarbon dates expressed in uncorrected radiocarbon years bp and bc/ad:

Table 3: Radiocarbon dates from Cashelkeelty I peat monolith.

Laboratory Number	Depth (cm)	C^{14} date bp	C^{14} date bc/ad
UB-2372	46-47	265 ± 95	1685 ad
UB-2371	115-116	1360 ± 85	590 ad
UB-2370	151-152	2470 ± 60	520 bc
UB-2414	183-186	3090 ± 60	1140 bc
UB-2368	215-218	3965 ± 60	2015 bc
UB-2367	255-256	4915 ± 95	2965 bc
UB-2413	291-294	5845 ± 100	3895 bc
UB-2366	299-300	4035 ± 85	2085 bc

It can be seen from this table that the date obtained for the basal 2 cm of peat (UB-2366) is too young and does not fit into the general sequence. A possible explanation for this phenomenon is that since the basal peat is resting directly on bedrock, seepage of water along the interface between the peat and rock faces of the hollow could result in the accumulation of more recent soluble humic acids in the base of the hollow, thereby contaminating the basal few cm. As will be seen later, the remaining dates of the sequence agree remarkably well with those suggested on the basis of vegetational changes.

The rate of formation of the Cashelkeelty I peat is illustrated in Fig. 10 where the radiocarbon dates obtained, are plotted against depth. This graph will be used to interpolate dates for features or events which occur between the radiocarbon dated samples. Such interpolated dates (denoted by * in the text) should merely be regarded as rough estimates, since some changes in formation rates of the peat are likely to have occurred between the samples dated by C^{14}. On the basis of extrapolation, the basal peat should date to c 4500 bc.

7.2.1.5 Vegetational History

Zone CK-1. c 4500*-2965 bc (zone VIIa, *sensu* Jessen; zone VII *sensu* Mitchell).

Zone CK-1 is one of considerable woodland cover where the tree pollen averages 68%. *Pinus*, *Betula* and *Quercus* form the major woodland components with *Corylus* present throughout but with relatively low values (average 5%). *Ulmus* is poorly represented as is usual in S Ireland diagrams (Jessen 1949) but reaches its absolute maximum of 1.9% in this zone. *Alnus* is present and is slowly expanding while *Ilex* appears towards the end of the zone at c 3190* bc. *Hedera* (ivy) pollen is abundant throughout the zone.

The local growth of *Salix* is demonstrated by the extremely high pollen values and the clusters of its pollen grains noted during analysis, which must represent catkins fallen onto the peat surface. No basal limnic sediments were found which indicates that the hollow never contained permanent open water. The peat-forming plants, as indicated by the pollen analysis were dominated by *Salix*, with the ground-cover vegetation consisting of *Sphagnum* mosses, Gramineae, *Potentilla* t. and *Polypodium* and *Dryopteris* (ferns). *Juncus effusus* (common rush) must also have been growing on the peat surface as evidenced by its seeds recovered during macrofossil analysis (Fig. 17).

There is evidence for some woodland disturbance and clearance during this zone. At 286 cm and 261 cm the total tree pollen values fall to 46% and 49% respectively, and some traces of charcoal were noted in the microscope preparations from 261 cm. The Gramineae (grasses) and *Pteridium* (bracken) curves peak at both levels and cereal-type pollen has been identified at 276 cm-291 cm (3895 bc). Using phase-contrast microscopy, and the criteria of Grohne (1957) and Beug (1961) these three cereal grains have been identified as *Hordeum* type (1 grain) and *Triticum* type (2 grains). Other indicators of woodland clearances and possible agricultural activity are the taxa normally occurring in pastures or arable and ruderal habitats. Of these, the highest frequencies are those of *Succisa*, Compositae lig. and Cruciferae, while there are fleeting appearances of *Plantago lanceolata*, *Ranunculus*, Leguminosae, Compositae tub. and *Stachys* t. (woundwort).

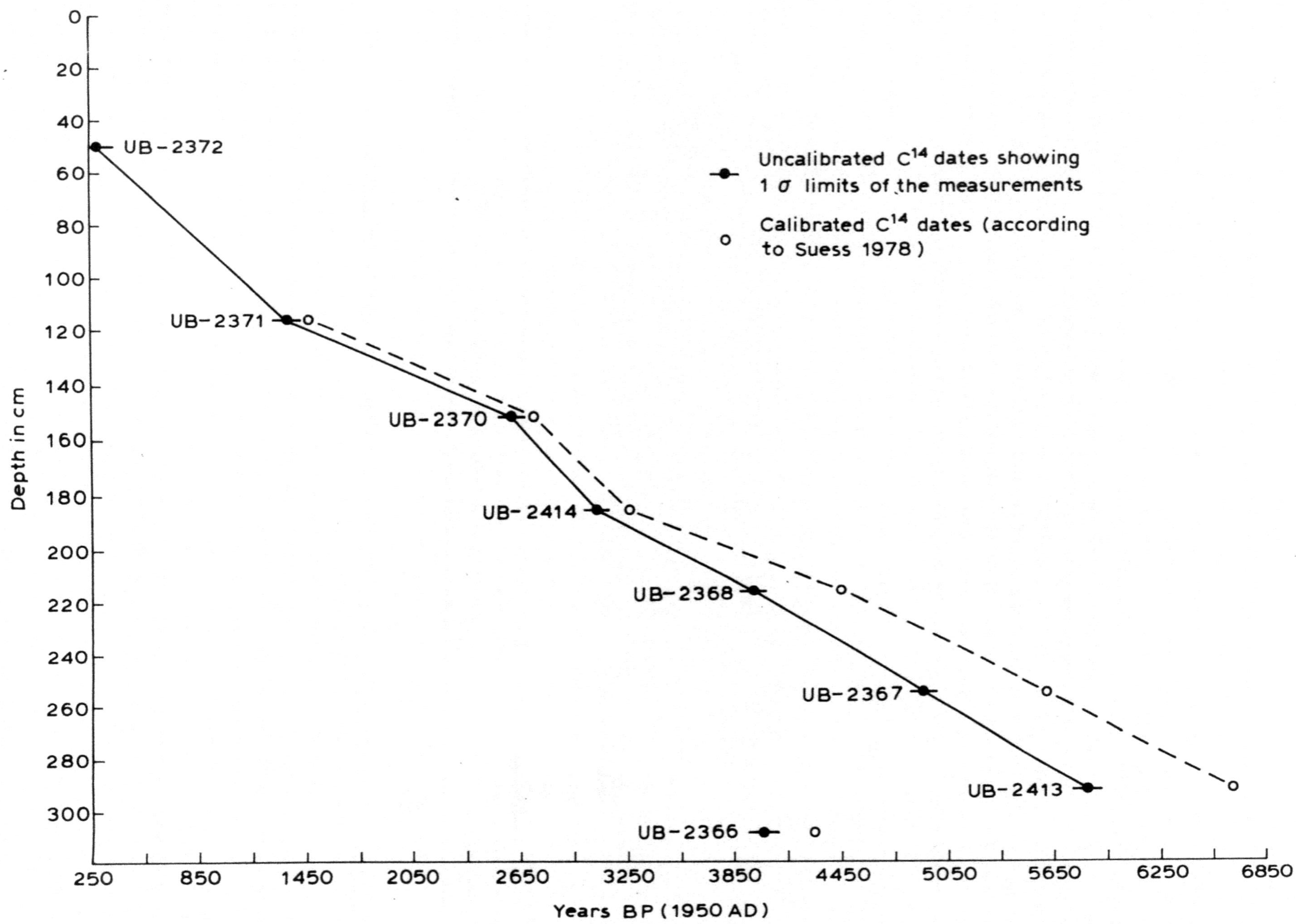

Fig. 10 Graph of Cashelkeelty I C^{14} dates plotted against depth.

The evidence of falling tree-pollen values, increasing grasses and the presence of charcoal, cereals and agriculturally significant weed taxa all point to human interference with the woodland cover and possible agricultural activity from early in the fourth millennium bc. This will be discussed fully below in section 7.2.5.

The correlation of the lower boundary of this zone with those of Jessen & Mitchell, is rather tenuous. Jessen (1949) accepts the rational border (i.e. where the pollen curve begins to rise to sustained high values) of Alnus as the boundary between his zones VI/VIIa (i.e. Boreal/Atlantic transition). Mitchell (1956) on the other hand maintains that as Alnus pollen values rise very slowly in Ireland, the rational border is inapplicable as a criterion for defining the zone boundary. Alternatively, his criteria fix the boundary where Pinus begins to fall, Alnus begins to rise and Corylus is usually at a minimum. Accepting these latter criteria for this diagram would place the Boreal/Atlantic transition at c 260 cm or c 3000* bc which is much too late. An almost identical situation was encountered by Pilcher at Slieve Gallion (1973, 685) and Ballynagilly (1979, 354) and as a result the zone boundary was drawn at the rational border for Alnus.

At Cashelkeelty I the continuous curve for Alnus begins at 290 cm (c 3790* bc) with a single grain in the basal layer (c 4500* bc). The lower zone boundary is drawn at the base of the deposit, just below what appears to be the rational border of Alnus (but in the absence of deeper deposits the exact location of empirical (i.e. point of arrival) and rational borders is uncertain). The work of Smith and Pilcher (1973) has shown that the empirical and rational limits of Alnus in N Ireland and Britain are clearly diachronous. It is postulated that Alnus arrived in the lowland areas of N Ireland c 6000 bc but did not spread to the upland areas until c 3500 bc (Pilcher, 1979, 355). If the beginning of the continuous curve for Alnus at Cashelkeelty I is accepted as its rational limit, then its date of 3790* bc is in general agreement with the lowland sites of N Ireland.

Zone CK-2. 2965 bc-c 1510* bc (part of Zone VIIb sensu Jessen, Zone VIIIa sensu Mitchell).

This zone is one of extensive woodland coverage with Quercus, Betula and Ilex being the major constituents. Alnus reaches its absolute maximum (36.9%) and Corylus maintains fairly constant values throughout. Pinus is declining rapidly and Ulmus never regenerates to its former values. Fraxinus (ash) appears for the first time at c 2440*bc, and two grains of Tilia (lime) were recorded. Hedera values attain a maximum of 4.7% during this zone.

At the lower boundary of the zone, Salix values drop rather sharply to 4% before again expanding to a peak of 124.9% at 246 cm from which point they gradually decrease towards the top of the zone. This temporary decline of Salix at the zone boundary corresponds with a peak in the Alnus curve and could perhaps be due to competition by Alnus trees which have spread onto the swampy peat surface. Salix however, re-establishes itself briefly before finally fading out at the upper zone boundary. The local ground-cover vegetation during this phase consists of Sphagnum cf. fallax moss, Juncus effusus, Juncus papillosum, Juncus articulatus (jointed rush) and Polypodium and

Dryopteris ferns. Ericaceae values are very low throughout and Gramineae are not present until the end of the zone.

The high tree pollen values (90-95%) point to almost total woodland regeneration during this zone. The absence of grasses until 216 cm (2015 bc) suggests that the clearings of Zone CK-1 had been abandoned and completely overgrown with scrub woodland. Apart from some scattered grains of Succisa and a peak of Cruciferae at 236 cm (c 2440* bc) there are no indicators of human activity. The re-appearance of grasses at 2015 bc followed shortly afterwards by Plantago lanceolata marks the onset of large-scale clearance in Zone CK-3.

Zone CK-2 has its lower boundary at the point where Ulmus values fall from 1.0% to 0.2% over 5 cm, Pinus reaches a maximum before beginning its final decline and Alnus reaches a peak of 20.5%. The Ulmus values are in general so low, that it is difficult to speak in terms of an 'elm decline' in the true sense of the term, but the behaviour of the Pinus and Alnus curves is characteristic for the period of the elm decline (Jessen,1949, 245; Smith 1961) and the sharp rise in Quercus values immediately afterwards is paralleled in Mitchell's observations for the same period (Mitchell,1956, 187). During Zone CK-1 (Atlantic period) the pollen of Hedera attained relatively high frequencies (max. 3.6%) but decreased sharply at the 'elm decline' before expanding again during Zone CK-2. A similar decline in Hedera pollen in the Danish diagrams has been used by Iversen to define the Atlantic/Sub-Boreal boundary (Iversen, 1944).

The decline in Ulmus values, such as it is, at the lower boundary of Zone CK-2 may therefore be correlated with the classical elm-decline of NW Europe which is accepted as the Atlantic/Sub-Boreal transition. The date of 2965 ± 95 bc relates to the end of, or just after the elm decline and as such, is in good agreement with those from similar contexts in N Ireland and Britain, all of which cluster around c 5000 bp (Smith & Pilcher, 1973, 910).

Zone CK-3. c 1510*-520 bc (part of Zone VIIb and VIII sensu Jessen and part of Zone VIIIb sensu Mitchell).

The lower boundary of zone CK-3 is defined as the point at which the total tree pollen begins to decline steadily. The boundary falls at c 1510* bc when Betula and Ilex have reached their final maxima before extensive clearance takes place. Quercus and Alnus are also affected by these clearances but Fraxinus which is not present early in the zone, is continuously present in the latter part (from c 940* bc) which could be due to its expansion onto newly-abandoned clearings. Ulmus values remain low and Hedera has virtually disappeared by the end of the zone. Corylus expands gradually, probably as scrub on the margins of the clearings. The Pinus curve finally wedges out at c 1040* bc which is younger than the youngest dates obtained for the final Pinus decline in N. Ireland, e.g. 1660 ± 75 (UB-376) from the blanket bog deposit at Glens Bridge, Co. Antrim (Smith and Pilcher, 1973, 908). Salix has dropped to values of c 1% and the transition to more oligotrophic conditions locally is evidenced by the appearance of plants characteristic of ombrogenous blanket peat vegetation. Ericaceae and Cyperaceae are expanding rapidly and Myrica gale and Potentilla t. have become well established.

Some local pools of open water are indicated by the presence of Eupotomageton (pondweed) pollen and the seeds of Potamogeton gramineus at 180 cm. Other local plants as indicated by macro-remains include Juncus effusus and Juncus articulatus, Ranunculus flammula (lesser spearwort), Carex acuta (slender-spiked sedge) which was probably growing along the edges of the pools and Eleocharis multicaulis (many-stemmed spike-rush). The ferns Polypodium and Dryopteris have declined together with the woodland.

This zone is one of intense human activity and corresponds with middle/late Bronze Age times. The woodland clearances allowed the expansion of Rosaceae, Gramineae and Pteridium. Agricultural activity is evident in the form of cereal cultivation throughout, with possible emphasis on pastoralism from c 760* bc as indicated by peaks in the curves of Plantago lanceolata, Ranunculus, Succisa pratensis and the presence of Polygalaceae (milkwort).

Zone CK-4. 520 bc-590 ad (Part of Zone VIII sensu Jessen and Zone VIIIb in part, sensu Mitchell).

This zone which covers the Iron Age period, begins when the total tree-pollen starts to rise gradually and there is a marked decline in human activity. This phase of woodland regeneration begins about 590 bc as Corylus scrub expands rapidly and invades the cleared areas. The pollen values for Betula, Fraxinus and Ilex also increase considerably while Quercus, Alnus, Salix and Ulmus remain fairly constant. We may therefore envisage the development of a more open type secondary woodland consisting chiefly of Corylus and Betula together with Fraxinus and Ilex.

The Gramineae curve falls steadily and even though a few cereal-type pollen grains are present, the pollen of weeds is virtually non-existant. Plantago lanceolata is present early in this zone but then disappears completely. Agricultural activity has obviously declined drastically since Bronze Age times with pastoralism and cultivation taking place in the area on an apparently small scale.

The vegetation of the blanket peat indicates a drying-out of the surface. Ericaceae values remain fairly constant but there is a marked fall in the curves of Cyperaceae, Sphagnum and Potentilla t. with a corresponding increase in the Myrica curve. The macrofossil evidence substantiates that of the pollen and shows a marked decline in the remains of rushes, sedges and mosses from c 670* bc. Drier conditions are also suggested by the stratigraphy of the peat throughout this zone which is relatively highly humified. The implication of these changes is discussed in section 7.2.5.

Zone CK-5: 590 ad-1685 ad (Zone VIII in part, sensu Jessen and Zone IX sensu Mitchell)

The lower boundary of this zone is placed where woodland clearance is again evident and agricultural activity intensifies. This occurs in Early Christian times and the zone covers the subsequent Viking and Medieval periods.

The curves for Corylus, Betula, Fraxinus and Ilex all show a general decline while Quercus and Alnus maintain fairly constant levels. Scattered grains of Fagus (beech) were identified.

The Gramineae curve rises to a peak of 74.9% at c 1400* ad and Plantago lanceolata is present throughout. Weeds of cultivation are well represented as are cereals, predominantly Triticum type (wheat). Avena type (oats) is present for the first time at the lower zone boundary (590 ad).

A return to wetter conditions on the bog surface is evidenced by a sharp rise in the curves for Cyperaceae and Sphagnum (papillosum ?). Narthecium ossifragum reaches values of 40% and the Ericaceae are also expanding, carbonised flowers of which were found during the macrofossil analysis. Rhynchospora alba (white beak-sedge) and Ranunculus lingua (great spearwort) are also growing locally. The Myrica curve has fallen to low levels as also have the curves for Pteridium and Dryopteris.

Zone CK-6. 1685 ad-present (Zone VIII in part, sensu Jessen and Zone X, sensu Mitchell).

The lower boundary of this zone coincides with the re-appearance of Pinus, just before a major intensification of agricultural activity in the area. The total tree-cover fluctuates, with a clearance phase at c 1710* ad when Quercus, Betula and Fraxinus were affected, but slight regeneration again took place when Quercus, Betula, Corylus, Fraxinus and particularly Ilex expanded, before the final decline to the present-day levels. It is interesting to note that even though a mature Pinus plantation stands c 150 m to the north of the sampling site, the Pinus values in the upper levels of the profile never exceed> 2%. Apart from the re-appearance of Pinus, little evidence of the extensive mixed-wood plantations established in the past two centuries in the neighbourhood (e.g. Derreen Wood and Derryconnary Wood) has registered on this pollen diagram which again emphasizes its restricted pollen catchment area.

The phase of intensive agricultural activity, which on the basis of interpolation covers the greater part of the 18th century, is marked by the woodland clearance just mentioned and by a sudden, sharp increase in the weed curves such as Plantago lanceolata, Ranunculus, Leguminosae, Rumex acetosa type (sorrel) and Urtica (nettle). Cereals (Triticum type) are also present. This evidence for extensive pastoralism and cultivation in an area which was by then largely peat-covered and with degraded soils, must represent considerable pressure on land resources, possibly associated with the pre-famine population expansion (see p. 97 below). The drastic reduction in population during and after the famine years (from c 1840 onward) allowed some secondary woodland regeneration in the abandoned farmlands but it was the blanket peat, with vegetation dominated by Ericaceae, Gramineae and Cyperaceae with some Myrica and Potentilla type, that came to dominate the area.

7.2.2 CASHELKEELTY II (Fig. 18, Table 8).

7.2.2.1 Site Description. This pollen diagram was prepared from samples taken during excavation from the sectional profile which extended from the five-stone circle to the eastern stone of the alignment (Fig. 7). This sectional profile was of considerable importance in establishing the temporal relationship between both monuments. The present-day local topographical and vegetational details are as described for Cashelkeelty I (see above p. 76).

7.2.2.2 Sampling and Stratigraphy. The 7 cm of peat were removed as a sod and sealed in polythene before removal to the laboratory. The remaining samples from the mineral subsoil were collected in open-ended glass test-tubes which were subsequently sealed.

The stratigraphy of the profile at the sampling point was as follows:

1-7	cm	Dry, brown peat of moderate humification. Visible root penetration in the upper 2 cm.
8-19	cm	Hard, grey compact stony layer which had been leached and gleyed. This layer was deposited during construction of the stone circle.
20-22	cm	Greasy, black highly organic layer which represents the old ground surface associated with the stone circle.
23-24	cm	Grey leached horizon in the boulder clay.
25-28.5	cm	Brown illuvial horizon in the boulder clay.

7.2.2.3 Presentation of Data. The pollen values were calculated as percentages of total pollen, excluding spores of ferns and other lower plants. Sampling intervals were irregular and respect the stratification. The pollen frequencies for the peat deposit are represented in the standard graph form while those from the mineral soil are represented as histograms and do not imply a temporal sequence. Only those taxa of relevance for the discussion are illustrated in the pollen diagram (Fig. 18) but the remainder are presented in Table 8.

7.2.2.4 Radiocarbon Dating. Three samples relating to this profile were dated at the Radiocarbon Laboratory, State University, Groningen.

Table 4 Radiocarbon dates relating to Cashelkeelty II

Laboratory Number	Context	C^{14} dates bp	C^{14} dates bc/ad
GrN-9171	Basal 2 cms of peat	800 ± 30	1150 ad
GrN-9172	Charcoal from the stony layer (G)	2665 ± 50	715 bc
GrN-9173	Charcoal from the old ground surface	2920 ± 60	970 bc

All samples received acid pretreatment and the whole fraction of the peat sample (GrN-9171) was dated. The charcoal samples (GrN-9172 and GrN-9173) were collected by careful hand-picking during excavation.

7.2.2.5 Vegetational History. Since the peat deposit is so shallow (7 cm), pollen assemblage zones have not been established for this profile. The pollen record from each stratigraphic horizon shall be considered separately and the reader is referred to chapter 4 above, where the problems inherent in the pollen analysis of mineral soils are discussed.

In general, pollen preservation in the mineral soil was quite poor. The pollen sum for the old ground surface is 57 which is much too low to provide a statistically reliable sample but some tentative suggestions may be made on the basis of the surviving pollen. The pollen sealed in this old ground surface were deposited in the few years immediately preceding the construction of the stone circle and should therefore give some indication of the vegetation cover of the Cashelkeelty area at this time. The tree pollen is dominated by Corylus, and Betula which suggests secondary woodland cover. Quercus must also have been growing in the locality as evidenced by its charcoal found embedded in the old ground surface and from which the C^{14} date was obtained. The high values for the ferns Polypodium and Dryopteris are most likely due to differential preservation, since their spores seem most resistant to decay and are usually found at depths where little or no pollen survives. Heathland conditions appear to have been well established in the area, as evidenced by the high Ericaceae values. Some grassland with Compositae lig. is also indicated.

The pollen in the overlying stony layer (G) which was deposited during construction of the stone circle, must be derived from the pollen rain of the period between the construction of the circle and the initiation of peat growth (i.e. the period 715 bc-1150 ad). This conclusion is based on the fact, that, since the strata from which the stony layer was derived (the $A_{o/1}$, A_2, B_2 horizons) contained little or no pollen, the grains surviving in the upper cm of the stony layer must have become incorporated after the layer was deposited and before it was sealed by the spreading blanket peat. Again, the picture presented is one of extensive heathland in the area with some secondary woodland comprising Corylus, Betula and Alnus. The charcoal found incorporated in the stony layer included fragments Corylus, Alnus and Salix. Some grassland is present and limited human activity is suggested by the single cereal grain and the presence of weed species including Plantago lanceolata, Compositae lig., Ranunculus, Leguminosae and Caryophyllaceae.

The overlying peat deposit provides the pollen record for the period 1150 ad to present, which corresponds with the later part of CK-5 and CK-6 of Cashelkeelty I. The curves for Corylus, Betula, Alnus, Ilex and Quercus remain fairly constant throughout and Pinus expands rapidly after c 1618* ad. The Gramineae curve falls sharply at c 1618* ad but again recovers and the frequencies of the weed taxa remain quite constant. The local vegetation is that characteristic of ombrogenous blanket peat in the area and includes high frequencies of Ericaceae together with Myrica gale, Potentilla t., Cyperaceae and Sphagnum.

7.2.3 CASHELKEELTY III (Fig. 18. Table 9)

7.2.3.1 Site Description. During excavation, a sectional profile was exposed which ran E/W along the centre line of the stone alignment and the pollen diagram has been prepared from samples taken from this profile at a point midway between the centre and westernmost stone (Fig. 7), and just 2 m distant from the sampling point of Cashelkeelty II.

7.2.3.2 Sampling and Stratigraphy. The sampling was carried out as for Cashelkeelty II, with the peat removed as a sod and the mineral soil collected in glass tubes.

The stratigraphy was recorded as follows:

1-11	cm	Dry, brown peat of moderate humification.
12-19	cm	Boulder clay upcast during construction of the stone alignment and subsequently leached and gleyed. Light grey in colour and highly cemented.
20-22	cm	Greasy, brown organic layer representing the old ground surface of the stone alignment.
23-24	cm	Grey, leached horizon in boulder clay.
25-45	cm	Brownish/orange illuvial horizon.

7.2.3.3 Presentation of Data. The pollen values are expressed as percentages of total pollen excluding the spores of ferns and other lower plants. As with Cashelkeelty II, the pollen frequencies of the mineral soil are represented as histograms while those of the peat are drawn as curves. The taxa not illustrated in the pollen diagram are presented in Table 9.

7.2.3.4 Radiocarbon Dating. The basal 2 cm of peat (whole fraction) from this profile was processed at the Groningen Laboratory:
GrN-9279, 1060 ± 30 bp (890 ad).

7.2.3.5 Vegetational History. Pollen assemblage zones have not been established for the shallow peat deposit and again, the pollen record of each stratigraphic horizon shall be considered separately.

Unlike Cashelkeelty II, the preservation of pollen in the top of the illuvial B_2 horizon was quite good and provides an indication of conditions prevalent during the early stages of the regional podzolization (see above p. 56), prior to the construction of the stone alignment. The high Ericaceae values (62.5%) indicate widespread heathland and local woodland or scrub regeneration is suggested by the presence of *Corylus*, *Betula* and *Alnus*. The rather low Gramineae values (2%) could mean that there was limited open grassland in the locality but the high frequency of *Succisa pratensis* suggests the presence of upland pastures. It is however, possible that the latter species was present as a component of the heath vegetation.

The picture presented by the old ground surface is very similar to that obtained at Cashelkeelty II, in spite of the better preservation of the pollen in this profile. Widespread heathland, some secondary woodland or scrub and limited grassland with few weed taxa are represented. Three cereal grains (cf. *Triticum* type) suggest cultivation in the locality.

The question of the origin of the pollen contained in the soil deposited on the old ground surface, which was subsequently gleyed, is of vital importance to the interpretation of the pollen spectra from this horizon. It is possible that some pollen was present in the soil when it was deposited (probably a mixture from the $A_{0/1}$, A_2 and B_2 horizon) and further pollen deposition took place on the soil surface when it was exposed between the time of construction of the alignment and peat initiation (a period of c 1800 years). The pollen record of the G horizon is therefore of little interpretative value since it is partly derived from older deposits and represents such a long period of time without a clear temporal sequence. It is interesting to note however, that the upper sample analysed in this deposit (at 15 cm) produced higher frequencies of *Pteridium* and weed taxa than the underlying horizons, frequencies which must be attributed to pollen deposition in the period between the construction of the alignment and peat initiation.

The pollen record provided by the peat, covers the period 890 ad to present. A phase of clearance and cultivation is seen at c 1010*ad when the curves for Corylus and Betula fall and those of Gramineae and Plantago lanceolata rise and a grain of Hordeum type (barley) was identified. The re-appearance of Pinus apparently occurs at c 1490*ad, although it is possible that this interpolated date is too early because of the less humified upper levels of the peat.

7.2.4 CASHELKEELTY IV (Fig. 19, Table 10)

7.2.4.1 Site Description. A monolith of peat and mineral soil was recovered from a point mid-way between the stone alignment and the multiple-stone circle (Fig. 7). The location of this sample was chosen so that it lay c 2 m east of a well-preserved stretch of pre-peat field fence, which, together with the excavated stretch running east/west, possibly forms one corner and side of a prehistoric field. It was hoped that pollen analysis of this monolith would provide evidence of the agricultural activities carried out within this field.

7.2.4.2 Sampling and Stratigraphy. Sampling was carried out using the metal boxes described above (p. 49).

The stratigraphy was recorded as follows:

1-14	cm	Dark brown peat of moderate humification.
15-17	cm	Mixture of sandy soil, small stones and a little peat. This is an erosion level.
18-22	cm	Highly humified, black greasy peat.
23-27	cm	Peat and mineral soil mixture.
28-34	cm	Gleyed boulder clay, grey and compact.
35-36	cm	Dark brown humus. A relict $A_{o/1}$ horizon of regional podzolization, on top of which lies a thin, undulating iron pan.
37-56	cm	Orange, illuvial (B_2) horizon.

7.2.4.3 Presentation of Data. The pollen values are expressed as percentages of total pollen excluding spores of ferns and other lower plants. Curves and histograms are used as in Cashelkeelty II and III. A hiatus is present at 15 cm where an erosion surface, possibly created by peat-cutting, is clearly indicated by a layer of sandy soil and small stones mixed with a little peat. The pollen spectrum from 26 cm is illustrated in histogram form since it is derived from a layer of peat and mineral soil mixture which probably represents a cultivation layer.

Because of the disturbed nature of the profile, no samples were submitted for radiocarbon dating.

7.2.4.4. Vegetational History. The spectra from the A_R and B_2 horizons together should provide information regarding conditions prevailing during the regional podzolization (see p. 54 above), before a change in the level of the water-table ultimately caused the formation of the iron pan with subsequent gleying of the overlying soil. The picture presented is one of considerable woodland cover, predominantly secondary woodland, comprising Corylus, Ilex and Betula but Quercus and Pinus and Alnus were also present. There is little evidence for open grassland and heath was poorly established.

The spectra from the gleyed horizon illustrate little change and must of course contain some pollen contemporary with the A_R and B_2 horizons.

The disturbed layer of peat and mineral soil shows a marked decline in Pinus values and also in those of Corylus and Ilex. The secondary woodland was apparently being subjected to clearance and there is a corresponding rise in the frequencies of grasses and several weed taxa which are suggestive of pastoral activities. The sudden increase in Ericaceae values and the appearance of Potentilla t and Cyperaceae must be attributed to the peat in this layer. No cereals or reliable indicators for arable farming were identified.

Pinus has just about disappeared in the base of the peat deposit proper (21-22 cm) which must place it in the period post-1040*bc (see p. 82 above). There has been a sharp expansion of grasses at the apparent expense of Corylus, Ilex and Betula and local wet conditions are evidenced by the rising Cyperaceae curve.

Above the erosion level, the vegetational history agrees with that already established for medieval/recent times at Cashelkeelty. Declining woodland cover, spreading bogland and the re-appearance of Pinus.

7.2.5 Discussion of the vegetational history of Cashelkeelty

7.2.5.1. Zone CK-1. c 4500*bc-2965 bc earlier Neolithic:
Peat accumulation started on the southern side of the Cashelkeelty ledge at c 4500*bc. This peat was restricted to a local swampy hollow and supported a Salix carr with a moss- and fern-rich understorey. There is little indication that this swampy hollow ever contained permanent open water and eventually, continued peat growth led to increased acidification and a decline in minerotrophic influences which, in turn, caused a transition from soligenous wood peat to ombrogenous blanket peat formation. Jessen's investigations in the bogs of Cork and Kerry, revealed many cases where a similar sequence of a wood peat vegetation gradually gave way to ombrogenous blanket peat. At nearby Toreen More West, Co. Cork the trees growing on the bog were chiefly Betula and Pinus while at Emlaghlea, Co. Kerry, Salix and Pinus were recorded (Jessen,1949,-189, 190). Other instances of blanket peat overlying wood peat have been recorded by Simmons (1964, 168) at Blacklane, Dartmoor and by Birks (1972, 198) at Snide Bog, Kirkcudbrightshire, Scotland.

The pollen flora of this period suggests that the Cashelkeelty landscape was heavily wooded. The high values for Pinus (maximum 40%) indicate that it must have been an important constituent of the woodland and it was most likely growing on the ledge itself and on the shallow acidic soils of the hillside. Betula was also a prominent tree and probably grew, together with Pinus, on the hillside. The low values for Corylus suggest that it formed the understorey in the woodland where its flowering was restricted by reduced light. If one accepts the conclusions of Janssen (1967) and Tinsley and Smith (1974) that the pollen representation of Quercus reflects its status in the surrounding vegetation reasonably well, then Quercus must have been well established in the Cashelkeelty woodland. The sharp increase of Quercus to 30% at 3895 bc with a corresponding decline of Betula and Pinus, may not be due to an actual expansion of the tree but rather to an increase in its relative contribution to the pollen rain as a result of clearance of Betula and Pinus.

Bearing in mind that Hedera helix is insect pollinated and produces pollen in low quantities, its high values throughout this zone suggest that Hedera was abundant on the Cashelkeelty ledge. The flowering of this evergreen (in October and November) is dependant on summer warmth and a mean temperature for the coldest winter month of not less than -1.5°C (Troels-Smith, 1960, 6) which points to mild climatic conditions throughout this period.

There is clear evidence for human interference with the woodland cover between 3895 bc and 2965 bc, during the early Neolithic period. The total tree pollen shows a sharp decline (from 70% to 46%) at c 3670*bc, after which regeneration takes place, and again from c 3400*-c 3070* bc there is a marked decline in woodland cover. The trees apparently affected by the clearance were Pinus and Betula and the possible use of fire for clearance is suggested by the presence of charcoal, noted in the microscopic preparations, at 261 cm or c 3070*bc. The decline in woodland cover is reflected by rising Gramineae and Pteridium values and the occasional appearance of weed and ruderal taxa including Plantago lanceolata, Ranunculus, Leguminosae, Succisa t., Rumex a.t., Compositae tub. and lig. and Cruciferae. Of particular interest are the three cereal-type pollen grains, one (Triticum type) at 291 cms (3895 bc) and two (Hordeum type and possible Triticum type) at 276 cm (c 3420*bc).

Taking all the evidence together therefore, it is clear that the Cashelkeelty area was occupied by man throughout the greater part of the fourth millennium bc. In view of the poor production and dispersal rate of cereal pollen, it is probable that some of the woodland clearance and cultivation was taking place on the mountain ledge itself. The relatively wide sample intervals (5 cm) do not allow a detailed reconstruction of the nature and duration of these Neolithic clearances but the following consideration of the available data provides an outline of events.

Two phases of activity are tentatively suggested. The first begins at 3895 ± 100 bc when values for Pinus and Betula are falling, Gramineae are beginning to rise steeply and weed taxa such as Leguminosae and Cruciferae are appearing. Cereal cultivation appears to have taken place immediately after clearance and the peaks in Gramineae and Leguminosae at 286 cm could be interpreted as evidence of grazing. It is interesting to note the isolated peak of Potentilla t., also at 286 cm, in view of Boulet's observation that Potentilla erecta is often associated with a relatively high grazing pressure (Boulet, 1939). Similar peaks in Potentilla t., at levels where other grassland/pastoral indicators reach high values, have been noted throughout this study. This first phase of activity appears to have ceased by c 3550*bc by which time woodland regeneration had taken place.

Within this first phase of c 400 years duration, it is impossible to say on present evidence, whether the events registered on the pollen diagram represent many small clearances in the neighbourhood or a few large scale disturbances of the woodland cover. The low values and sporadic occurrences of the weed and ruderal taxa, do suggest that local clearances were on a small scale and possibly short-lived. On the other hand, the possible marginal location of the sampling site should be kept in mind and considerable activity, along the coastal plain, for instance, could be responsible for the pollen of clearance and agricultural indicators. It is even possible that upwelling air-currents transported the cereal pollen grains from lower levels to the mountain ledge (cf. Oldfield,1976). The method employed to clear the woodland is not clear, but a combination of fire, felling by axe and ring-barking is on the basis of modern ethnographic parallels (Boserup, 1965, 24), most likely.

The second proposed phase of activity began at c 3420* bc when there is evidence of renewed clearance, primarily of _Pinus_ and _Betula_. The cereal grains again occur right at the beginning of the phase, and coincide with a high peak of Cruciferae. The curves for Gramineae, _Succisa_ t. and _Potentilla_ t. reach maxima at 271 cm (c 3300* bc), before finally disappearing at 2965 bc when woodland regeneration is well established. As with the earlier phase, the nature, extent and location of this human activity, is not clear, but while bearing in mind the inaccuracies inherent in interpolated dates, it does seem as if both land-clearance phases were of similar duration i.e. c 350-400 years. It is interesting to note that on the basis of radiocarbon dated pollen diagrams from N Ireland, Pilcher _et al_. (1971, 560) suggest Neolithic clearance phases lasting some 400-700 years.

The archaeological implications of early Neolithic activity, including cereal cultivation, in the Cashelkeelty area, are considered in the following chapter.

7.2.5.2 Zone CK-2. 2965 bc-c 1510*bc. later Neolithic/early Bronze Age: Throughout this zone, peat accumulation was still restricted to the swampy hollow on the southern side of the ledge and _Salix_ appears to have been competing with rapidly expanding _Alnus_, which towards the end of the zone, had become dominant. A fern-rich understorey with some _Sphagnum_ mosses was still present. The sharp temporary decline of _Salix_ at the lower zone boundary may have been caused, in part, by an expansion of _Alnus_ but the presence of microscopic charcoal fragments in the profile at this level could suggest some local clearance by fire. This point will be returned to shortly when discussing the elm decline. The presence of streaks of charcoal noted in the profile at 235-241 cm, corresponds with the final, sharp decline of _Salix_, but whether or not the latter phenomenon may be due to burning, is impossible to say on present evidence.

It has already been noted (see above p. 82) that the decline in _Ulmus_ values from 1.0% to 0.2% over 5 cm, just before 2965 bc may be equated with the elm decline which occurs with striking synchroneity at c 5000 bp throughout NW Europe. The nature and possible causes of this decline in _Ulmus_ pollen throughout NW Europe has been the subject of much discussion (e.g. Smith 1961; Ten Hove 1968; Sims 1973) and it is not intended to evaluate here, the various opinions presented. The various causes put forward range from climate (Iversen 1960), disease, pollarding or unfavourable ecological conditions (Ten Hove, 1968), to human interference in the form of felling, ring-barking or collection of the leaves and branches for cattle fodder (Troels Smith,1954; Mitchell,1956; Sims, 1973). Several aspects of the elm decline as recorded in Cashelkeelty I, are worthy of mention in relation to the above-mentioned causal factors.

The time-scale of the elm decline, as established from the point (261 cm) where pollen values fall from 1% to 0.2% (at 256 cm), is bracketed by the dates of c 3040* bc-2965 bc which means that a period of c 75 years is involved. There are marked changes in the general vegetation cover at this time: a pronounced rise in the curves of *Alnus* (also noted by Mitchell, 1951) and *Pinus* and less so in the curves of *Ilex* and *Betula*; a slight decline in the values of *Quercus* and *Salix* and the disappearance of the Rosaceae and *Hedera*; Gramineae are rapidly declining and the Cruciferae curve rises to a final peak while other non-tree pollen are either reduced to very low values (e.g. *Pteridium* and Cyperaceae) or have disappeared entirely (e.g. the weed and ruderal taxa).

The argument for a climatic cause of the elm decline is usually based on changes in the pollen frequencies of *Hedera*, *Ilex*, *Ulmus* and *Viscum* (Iversen 1944). The virtual disappearance of *Hedera* at the period of the elm decline in Cashelkeelty is quite spectacular, when viewed against the high values recorded throughout Zone CK-1 and later in Zone CK-2 after it re-appears at c 2800* bc. This could suggest a period of more continental climate with harsh winters, which would also have been unfavourable to *Ulmus*. The rising *Ilex* values however, militate against this view since according to Iversen (1944), *Ilex* is even more sensitive to cold winters than ivy and can hardly tolerate temperatures of less than 0.5°C.

To invoke human activity, whether in the form of felling, ring-barking or collection of leaves and branches for fodder in order to explain the Cashelkeelty elm decline, is hardly feasible. Throughout the elm decline period, the area of grassland was rapidly decreasing, being invaded by secondary woodland comprising *Betula*, *Pinus* and *Ilex* and apart from the Brassicaceae, all taxa related to agricultural activities were absent. It is therefore difficult to envisage large-scale selective clearance of elm trees as suggested by Troels Smith (1954) and Mitchell (1956) or collection of elm leaves for fodder, without other indications of such activities registering on the pollen diagram. The microscopic charcoal fragments at the beginning of the elm decline (261 cm) are the only possible indicators of interference with the vegetation cover at this time. Also, how is one to explain the apparent non-interference with *Ulmus* in the preceding phase of agricultural activity?

In an attempt to explain a similar decline in elm values without evident human activity, in N Scotland, Pennington *et al*. (1972) suggested that the pollen frequencies were affected by a general decline in elm resulting from human activity, further south. The restricted pollen catchment area of Cashelkeelty however renders such an explanation unlikely.

Finally, the possibility of an epidemic fatal disease of elm should not be disregarded. Widespread destruction of the elm trees of W Europe has resulted from the Dutch elm disease (caused by the fungus *Ceratocystis ulmi*) which was first recognized in the Netherlands in 1919. Serious epidemics have also been recorded in the U.S. during the late nineteenth century, caused by a virus disease carried by the incadelide *Scaphoides luteola*. These two diseases have however, been dismissed by Heybroek (1963, 3) as possible causes of the prehistoric elm decline, on the grounds that the remaining elm populations should have developed a high level of resistance to the diseases, which is not the case. Is it not possible however that a disease, to which the elm is now resistant, could have been the cause?

As Smith (1961a, 47) points out, the factors responsible for the vegetational changes at the time of the elm decline, may well have varied from region to region and were influenced by local conditions of soil, microclimate and ecological competition. It is suggested here that human activity cannot account for the decline in elm values at Cashelkeelty but rather disease and perhaps micro-climatic changes may have been the influencing factors. The low values of Ulmus to start with, suggest that it was not well established in the area and was more than likely restricted to the more base-rich soils of the coastal area. Once reduced therefore, (for whichever reason) competition from other trees, particularly the rapidly-expanding Quercus, Betula, Ilex, would have prevented its regeneration. Apart from a slight resurgence at 226-241 cm, Ulmus is absent throughout Zone CK-2.

The period 2965 bc-2050* bc was one of dense woodland cover with total tree-pollen rising to 97%. The high Quercus values (max. 63.3%) indicate its predominance in the locality, together with Betula and Ilex. The gradual decline of Pinus is probably due to its inability to compete with the species of the mixed oak woodland and by the end of the zone (c 1510* bc), has virtually disappeared. Corylus has increased slightly and was still probably restricted to the understorey layer. The rising Alnus values (max. 36.9%) may in part be due to its invasion of the swampy peat filled hollow but also indicates increasing oceanic conditions with consequent rising water-table. The woodland cover throughout this period (i.e. 2965-c 2050* bc) must have been dense with little or no clearings surviving as indicated by the absence of Gramineae pollen and that of weed taxa. There is no evidence to suggest that man was present in the Cashelkeelty area at this time, i.e. during later Neolithic times.

At c 2050* bc, the tree-pollen fell to 80% and Gramineae re-appeared suggesting some open ground. This idea is supported by the fleeting appearance of the Rosaceae, Plantago lanceolata and the re-appearance of Pteridium. Traces of charcoal were noted in the profile at this level (c 2050* bc) and it seems that some attempt was made to clear a little woodland (possibly by using fire) but Betula and Ilex rapidly regenerated before the large-scale clearances of Zone CK-3 began. Perhaps the most striking vegetational change during this stage of incipient forest clearance is the sudden fall in Hedera values from 2% to zero at 2015 bc and its absence until c 1510* bc. Ilex values also fall at 2015 bc but rise again immediately. It may be that a short period of climatic deterioration with cold winters occurred at c 2015 bc, with Ilex recovering and regenerating more rapidly than Hedera.

7.2.5.3 Zone CK-3. c 1510*bc-520 bc. middle/late Bronze Age:

The soligenous wood peat was gradually replaced by blanket peat with its characteristic ombrogenous vegetation including Ericaceae, Myrica, Potentilla t. and Cyperaceae. This transition from eutrophic to meso/oligotrophic conditions was a gradual one but seems to have been achieved by c 1100* bc. Podzolization and gleying was taking place on the mountain ledge and it is likely that by the end of the zone, blanket peat had spread onto the central part of the ledge, possibly as far as Cashelkeelty IV (see

p. 89 above). It is probable that the high Gramineae peak (51.4%) at the upper zone boundary is largely due to increasing contribution of grasses (e.g. Molinia) growing on the peat surface.

The total tree pollen declines rapidly from 90% to 29% between c 1510* bc and 520 bc with just one slight regeneration episode at c 850*bc. All constituents of the mixed oak forest were affected by these clearances; Quercus and Betula values gradually decline and by the end of the zone Ilex has virtually disappeared. Alnus maintains fairly constant values but must have suffered some clearance as evidenced by its charcoal noted in the profile at 190 cm. The Pinus curve shows a slight resurgence before finally wedging out at c 1040*bc. Identification of Pinus charcoal at 300 cm (c 1510*bc) indicates that the few trees remaining in the locality did not escape this clearance onslaught, which must have hastened the already established trend towards extinction through natural succession. The rising Corylus values after c 1040*bc and the re-appearance of Fraxinus could be due to scrub regeneration in abandoned clearings.

The sampling intervals used throughout this zone (2 cm and 5 cm) do not allow for a detailed sub-division on the basis of changing agricultural emphasis but the general sequence of events in the Cashelkeelty area throughout middle/late Bronze Age times may be reconstructed. As already noted in relation to the Neolithic activity (p. 90), the degree of resolution of anthropogenic events is such that evidence of several small-scale and short-lived clearances in the area may be masked by the general indications of clearance registering on the pollen diagram. The very nature of the forest fallow system of land-use implies that within the exploited area, clearings under cultivation, clearings lying fallow in varying stages of regeneration and untouched virgin forest, may all exist side by side. One should envisage such a mosaic of land-use patterns in the Cashelkeelty area during middle/late Bronze Age times.

Coincident with the evidence of woodland clearance (200 cm), the Gramineae curve rises sharply and cereals (Triticum type) appear, even though the weed taxa, excepting Leguminosae, have not yet become established. It therefore appears that once the trees in a plot were felled and/or burnt, the sowing of seeds and perhaps planting of roots,was carried out immediately in order to take advantage of the humus-rich soil with possible enrichment by ashes. The initially high yields obtained from such a plot could not have lasted for more than a few years however (Boserup, 1965, 24) and so new plots were prepared and the old left to provide rough grazing and grow secondary forest. The appearance of continuous curves for pastoral indicators such as Plantago lanceolata and Ranunculus may be dated to c 1140*bc and Potentilla t. has also re-appeared by this time.

It was towards the end of this phase of well-developed mixed farming when activities included the cultivation of Triticum (wheat) and pastoralism that the stone alignment was built at c 1000*bc. The pollen spectrum from the old ground surface related to the alignment (p. 87), presents a similar picture to that just outlined, with evidence of mixed farming activities in a local environment of spreading heathland and Corylus scrub.

A slight lull in clearance activity is apparent between c 1030* bc and c 850*bc, when rising curves of Corylus, Quercus and Alnus and the re-appearance of Fraxinus all suggest woodland regeneration. The Gramineae values decline as do the majority of weed taxa. The construction of the five-stone circle which isbracketed by the dates970 ± 60 bc and 715 ± 50 bc may well have taken place during this period when much of the Cashelkeelty hill-side must have been lying fallow. It is difficult to determine the real nature and cause of this lull in agricultural activity. It is possible that the preceding centuries of cultivation had resulted in widespread soil degeneration (as evidenced by the spreading heathland conditions) which may have caused a movement of the farming communities to a new area, or alternatively, emphasis may have shifted to exploitation of lands outside the pollen catch-ment area.

Renewed clearance and cultivation is evident at c 850*bc and continues right through to 520 bc. Hordeum type (barley) was now being cultivated in addition to Triticum type and the sharp peaks of Plantago lanceolata, Ranunculus and Potentilla t. at c 670*bc suggest a shift to predominantly pastoral farming. By 520 bc both pastoral and arable farming had apparently been reduced to a negligible level in the Cashelkeelty area. In the absence of radiocarbon dating it is impossible to correlate precisely, the period of cultivation which is represented in the basal peat at Cashelkeelty IV (see above p. 89) with any of the activity phases established from the other profiles. On the basis of the tree pollen contained in the Cashelkeelty IV cultivation layer however, (i.e. dominant Corylus, Ilex, Betula with a little Pinus) the earlier part of CK-3 would seem the most probable area of correlation.

7.2.5.4 Zone CK-4. 520 bc-590 ad. Iron Age/Early Christian:
A change to highly humified peat coincides with the early stages of Zone CK-4 and since the process of humification needs an aerated water medium (cf. Walker 1970, 121), it is suggested that the water table has ceased to rise or may even have fallen slightly. A drying out of the peat surface is also indicated by changes in the bog vegetation: the Sphagnum and Cyperaceae curves show a marked fall and there is a corresponding increase in the values of Myrica. The causes of such a dry phase are uncertain and could perhaps have been due to a general improvement in climatic conditions which began in this zone and culminated in the secondary 'climatic optimum' at c 1000 ad (see p. 5 above). Alternatively a fall in the water table level could be explained by attempted drainage on the mountain ledge. No evidence was recovered during excavation however, which would suggest that the Bronze Age farmers ever attempted to drain the peaty surface and if such was the case, then it seems strange that once their aim had been achieved (i.e. a drier soil) the area was apparently abandoned!

A vigorous growth in secondary woodland is envisaged between 520 bc and c 350*ad (i.e. Iron Age times) when Corylus, Betula, Ilex and Fraxinus apparently took advantage of the abandoned clearings which still remained free of peat. A little Quercus still survived in the neighbourhood but Alnus had reduced considerably.

Apart from the pastoral activity suggested by the peaks of Gramineae, Potentilla t. and Plantago lanceolata at 141 cm (c 100*bc) and the few cereals (Triticum type) at 131 cm (c 180*ad), there is little evidence of human activity.

There is however, a suggestion of renewed attacks on the secondary woodland at c. 350 ad when the values for Corylus and particularly Betula and Ilex decline appreciably, but indicators of farming activities are limited to a single grain of Triticum and rising Potentilla t. which can hardly be considered evidence of well-established agriculture. The final disappearance of Ulmus and Fraxinus which has been accepted by Watts (1961) and Mitchell (1965, 128; 1976, 166) as heralding the dramatic expansion of agriculture associated with the early monastic movement, has been dated to c 300 ad (McAulay and Watts 1961, 31). At Cashelkeelty, the Ulmus values rarely rose above 0.5% after 3000 bc and continued to appear sporadically right through to recent times which makes it impossible to pinpoint a time of final disappearance. The Fraxinus values are slightly higher but equally erratic and again, continue to recent times. The impact of Early Christian population increase and new farming practices as recorded by Mitchell at Littleton, Co. Tipperary (Mitchell, 1965) and at Redbog, Co. Louth (Mitchell, 1976) does not register on the Cashelkeelty pollen record with any great intensity until c 590 ad, which once again serves to stress the restricted pollen catchment area and local nature of the pollen record.

7.2.5.5 Zones CK-5 and CK-6. 590 ad to present. Early Christian to recent times:

A return to relatively wet conditions on the bog surface is evidenced by the resurgence of Sphagnum and Cyperaceae values. According to Tallis (1964), Sphagnum spore peaks tend to occur in positions where the peat stratigraphy indicates a sudden change in environmental conditions, and at Cashelkeelty, such peaks occur just after a change-over to a slower rate of humification implying a rising water table. Narthecium ossifragum appears to have dominated the peat vegetation throughout Zone CK-5 but gave way to Ericaceae in Zone CK-6. The Myrica curve has declined and the species is most likely restricted to the edges of trackways or the drier sections of the bog, as is the case today. The peat cover on the mountain ledge gradually edged outwards and covered the better drained elevations on which the stone alignment and circle were sited, during these zones.

Following the Early Christian woodland clearances, the total tree pollen curve shows slight fluctuations but with little change in the general composition of the woodland. Remnants of the mixed-oak woodland still remained, probably on the lower hill slopes and the coastal plain while Corylus scrub still occupied much of the abandoned farmland. Some Alnus must have been present in the wetter areas.

Avena type (oats) cereal appears for the first time at 590 ad and since the presence of this cereal is well documented in Iron Age Britain (Jessen and Helbaek, 1944; Dennell, 1976) its introduction to SW Ireland may perhaps be attributed to Early Christian missionaries. A slight intensification of both arable and pastoral farming is indicated at c 860*ad which corresponds with the arrival of the Vikings in Ireland. It is unlikely that Viking incursions or settlement took place west of Cork city but it may be that political unrest to the east, caused a certain amount of population movement to the less attractive western peninsulas, with resultant occupation of the more marginal lands. Similarly, the Norman conquest of Ireland in the early twelfth century had little or no effect on the southwestern corner of the country and the sporadic indications of agricultural activity which register on the Cashelkeelty diagrams up until 1685 ad must be attributed to the native Gaelic farmers in the area.

A further demonstration of the marginal location of the Cashelkeelty area with regard to the rest of Ireland, is the obvious escape of its woodland from the general devastation of Irish oak-woods during the late sixteenth, seventeenth and early eighteenth centuries. Throughout this period (1585-c 1770 ad), following on the Tudor plantations, the woods of Munster were continuously exploited and vast quantities of timber were exported for shipbuilding and barrel staves and what wood remained was used as charcoal for iron and glass works (Mc Cracken, 1971). This widespread devastation of Irish woodland registers clearly on the pollen diagrams from central and E Ireland, e.g. Littleton, Co. Tipperary (Mitchell, 1965) and Redbog, Co. Louth (Mitchell, 1976).

The final episode of clearance, cultivation and widespread pastoralism covers the greater part of the eighteenth century (26-41 cm). _Quercus_, _Betula_ and to a lesser extent _Corylus_ suffered some clearance but in general, it seems as if few new clearings were made and instead the peat-covered hillside was subjected to intensive agricultural activity. Dramatic peaks in the curves of _Plantago lanceolata_, _Ranunculus_, _Rumex acetosella_ t., and _Potentilla_ t. indicate that pastoralism was predominant, but _Triticum_ t. and _Hordeum_ t. were also cultivated. This sudden shift to the farming of wet, peaty, marginal lands must represent severe pressure on land resources in the area. After the Cromwellian wars of the mid-seventeenth century, the Irish landowners were forced to abandon their farms to the English planters and move west of the river Shannon to the wet, infertile areas. The Cashelkeelty area must have received its share of displaced Irish at this time, which, when combined with the rapidly rising population of the eighteenth century (between 1700 and 1840, the population of Ireland had risen from c two million to over eight million), serves to explain the need to eke a living from what would otherwise be classed as barren land.

This agricultural activity was short-lived however and when successive failures of the potato crop in the 1840s caused the Great Famine (1845-1851), thousands died and many of the survivors emigrated. By 1930, the continuing emigration particularly from the poorer western seaboard counties, had brought the population of Ireland down to about four million. The cabins on the Cashelkeelty hillside were abandoned and many of their ruins are still visible today, frequented only by sheep.

7.3 Dromatouk I, Co. Kerry (Fig. 20, Table 11)

7.3.1 Site Description

Townland: Dromatouk Altitude: 107 m OD
County: Kerry Nat. Grid. Ref.: VO95 071

A single monolith of blanket peat and mineral soil was taken from a point c 5 m west of the alignment of three standing stones at Dromatouk (Fig. 11). This site is located c 5 km east of Kenmare town (and c 25 km east of Cashelkeelty) on the northern 'shoulder' of Curramore mountain (235 m OD) which falls steeply to the valley of the Roughty river c 2.5 km to the north. The situation at Dromatouk is very reminiscent of that at

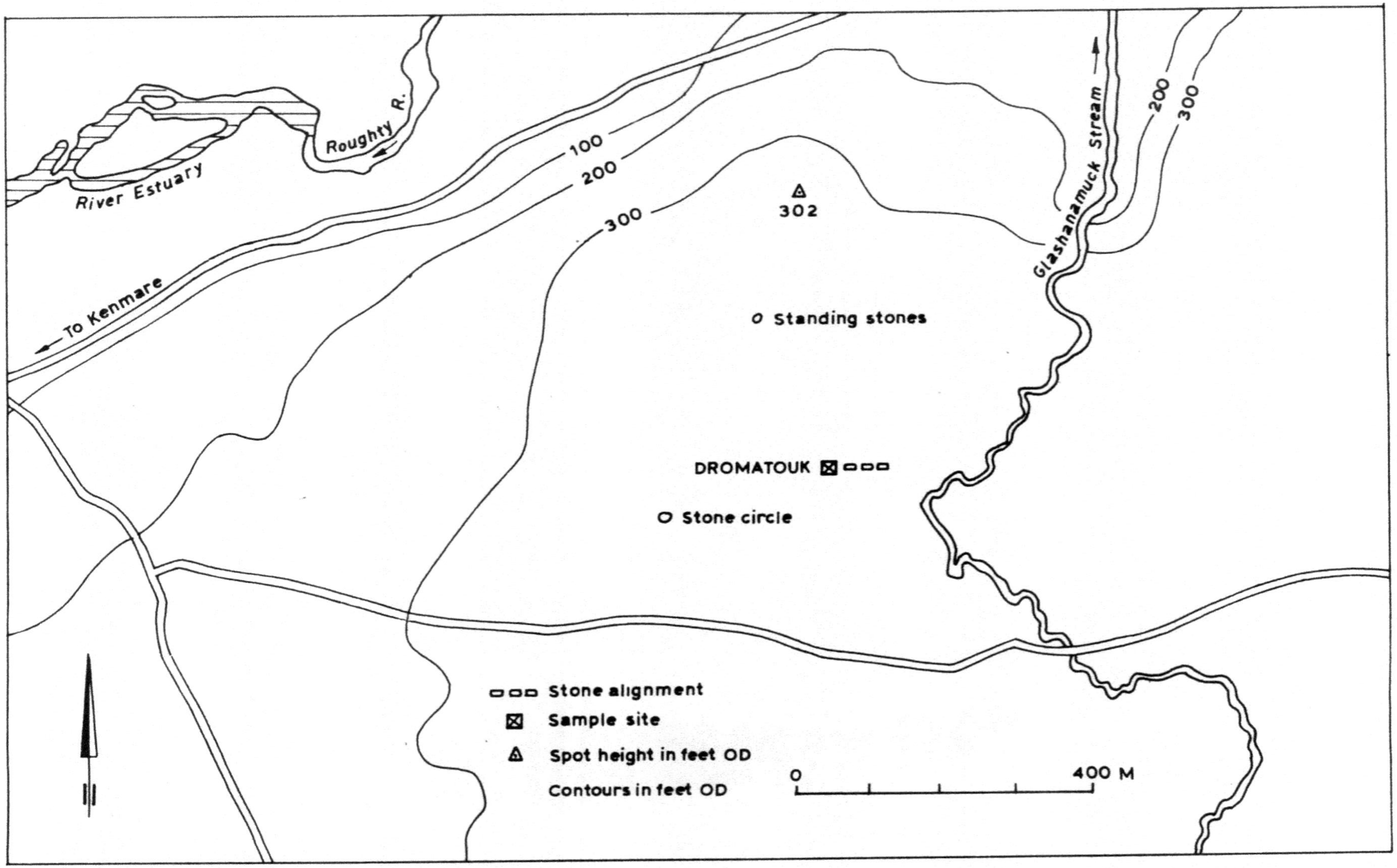

Fig. 11 Location of Dromatouk alignment and sample site.

Cashelkeelty, with the alignment itself built on a slight elevation in an otherwise level stretch of hillside, the greater part of which is covered with blanket peat. A stone circle is located c 2 km to the west and 1.7 km to the north are three standing stones one of which bears an ogam inscription.

The pollen catchment area of the Dromatouk diagram is likely to include the valley of the Roughty river in addition to the immediate hillside. As with the Cashelkeelty diagrams, the amount of regional pollen reaching the site, is not likely to be large because the Caha mountains to the south, the Derrynasagart mountains to the east, Mangerton mountain to the north and Kenmare estuary to the west, virtually enclose the site.

The local vegetation today is virtually treeless with great expanses of blanket peat supporting much _Calluna vulgaris_ (Ling), _Myrica gale_, _Narthecium ossifragum_ and _Potentilla erecta_. A little mixed farming is carried out on the sheltered valley side c 2 km to the north but the peat-covered hillside is used only for extensive sheep grazing.

7.3.2. _Sampling and Stratigraphy_. A pit (c 1 m^3) was dug 5 m west of the stone alignment and a monolith taken from the exposed face. Even on the crest of the elevation, the peat cover was c 40 cm deep and details of the stratification are as follows:

1-10 cm	Light brown, dry fibrous peat.
11-33 cm	Dark brown/black peat with some root penetration.
34-48 cm	Highly humified black greasy peat.
49-57 cm	Grey, gleyed boulder clay with many charcoal lumps occurring between 51 and 53 cm.
58-73 cm	Brown, illuvial horizon in the boulder clay.

7.3.3 _Presentation of Data_. The pollen values were all calculated as percentages of total pollen, excluding spores. Sampling intervals in the peat were generally 5 cm and 3 cm in the mineral soil. The results are presented in a pollen diagram (Fig. 20) and Table 11 and local pollen assemblage zones have been established for the peat spectra.

7.3.4 _Radiocarbon Dating_. Two radiocarbon dates have been obtained for this profile from the Groningen Laboratory.

Table 5 Radiocarbon dates relating to Dromatouk I.

Laboratory Number	Context	C^{14}date bp	C^{14} date bc/ad
GrN-9174	Basal 2 cm of peat	2020±30	70 bc
GrN-9346	Charcoal, 3-5 cm below surface of mineral soil	3330±50	1380 bc

Both samples received acid pretreatment and the whole peat fraction was dated.

7.3.5 Vegetational History

7.3.5.1 The Pre-Peat Vegetation: For reasons outlined on p. 44 above, the three spectra from the mineral soil shall be discussed as one. In fact the frequencies of species in each spectrum are almost identical, with the exception of the high (59%) values for *Pteridium* in the illuvial B_2 horizon, which may be due to the higher resistence to decay of its spores. The mineral soil spectra should provide a general picture of conditions prevailing during the acidification/podzolization of the soil prior to peat formation.

The average total tree pollen of 27% suggests, either that woodland existed some distance from the site (> 100 m) or that the site was surrounded by dispersed trees (cf. Tinsley and Smith 1974, 563). If woodland was present in the locality, it must have been rather open, comprising chiefly of the light demanding species of *Pinus*, *Betula* and *Corylus* with some *Ilex* but very little *Quercus*. The *Alnus* trees may have been a woodland component or growing along the banks of the nearby Glashanemuck stream. It is possible that much of the woodland was restricted to the Roughty valley, north of the site itself.

The degeneration/acidification of the local soils is evidenced by the high Ericaceae values and much of the hillside must have been covered by heathland throughout this period. Limited grassland in the area is indicated by the relatively low Gramineae values and apart from scattered grains of *Plantago lanceolata*, *Ranunculus*, Leguminosae and Compositae tub. /lig. there is little evidence for human activity.

The basal peat date of 2020±30 bp (GrN-9174) or 70 bc, provides a *terminus ante quem* for the vegetational conditions just outlined. The presence of *Pinus* in the samples suggests that acid soil conditions suitable for pollen preservation had developed by at least c 1100 bc (i.e. the date by which *Pinus* had disappeared from the pollen record in SW Ireland). In the absence of excavation of the nearby alignment, it is impossible to associate the charcoal layer in the mineral soil, with any particular phase of human activity. By analogy with the excavated alignments at Cashelkeelty and Maughanasilly, where the upper few cm of the gleyed (G) horizon were shown to be the activity levels associated with the sites, it is tentatively suggested here that a similar correlation may exist at Dromatouk. If such is the case, then the C^{14} date of 3330 ± 50 bp (GrN-9346) or 1380 bc obtained from the charcoal, could possibly be related to the construction phase of the alignment. Such a date would be in general agreement with that obtained at Maughanasilly (p. 71) and slightly earlier than that suggested for the Cashelkeelty alignment (c 1000 bc).

It is of interest to note that the charcoal, which has been identified as *Pinus sylvestris*, occurs at a level where the *Pinus* pollen values are extremely low. It could be that some clearance of *Pinus* trees in the locality was undertaken at the time of construction of the alignment, but this is of course, merely speculation.

The absence of indicators of human activity in the mineral soil could mean one of two things: either that the pollen of weed taxa and cereals did not survive or alternatively, agricultural activity was taking place some distance from the site with little of its pollen registering on the diagram. In view of the preservation, under similar conditions, of weed taxa pollen and cereal pollen at other sites including Maughanasilly I and Dromteewakeen I, it is unlikely that differential decay is the causal factor. It is far more probable that the area of settlement and agricultural activity of the alignment builders was sited some distance away—perhaps in the Roughty valley?

7.3.5.2 Zone DT-1. The podzolization process and subsequent gleying, were taking place for c 1000 years at Dromatouk before conditions became sufficiently waterlogged at 2020 ± 30 bp or 70 bc for peat formation to start. The local peat vegetation of this zone was that characteristic of ombrogenous blanket peat with Ericaceae dominant, high *Myrica* values and low *Potentilla* t., Cyperaceae and *Sphagnum*. The high frequencies of *Dryopteris* spores suggest that this fern was widespread on the heathy hillside.

The upper boundary of this zone is drawn where *Corylus* begins to decline and the weed taxa have been reduced to rather low values. There appears to have been little change in the woodland cover since pre-peat times, except that *Pinus* has virtually disappeared. *Corylus* and *Betula* are dominant amongst the tree species but *Quercus* and *Alnus* are also present.

The curves for Gramineae, *Ranunculus*, *Plantago lanceolata* and Leguminosae all rise to peaks at 46 cm which should date to shortly after 70 bc. This implies a short-lived period of pastoralism in the locality but no evidence of arable farming was recorded. It is possible that the hillside was used for sheep or cattle grazing for some time during the middle Iron Age—a period when similar activities are postulated for Cashelkeelty.

7.3.5.3 Zone DT-2. The upper boundary of this zone was drawn where evidence of intensive agricultural activity first appears (16 cm). Zone DT-2 therefore covers a long period of time, probably from late Iron Age/Early Christian times to the eighteenth century. It was not attempted to interpolate dates for this profile because between the dated basal peat layer and the surface sample, changes in rates of formation are suggested by the differing degrees of humification.

The local peat vegetation is still dominated by Ericaceae but *Myrica* values have been reduced to zero by 30 cm and the rising Cyperaceae curve suggests slightly wetter conditions. The only changes suggested for the woodland cover are a decline in *Corylus*, apparently due to encroaching bogland and a slight increase in the *Alnus* curve, all of which indicate increasing wetness. Indicators of human activity are limited to peaks in the pasture weeds *Ranunculus* and representatives of the genus Leguminosae at 25 cm and the rising values of *Plantago lanceolata* above that level.

It seems reasonable to conclude that the Dromatouk hillside was marginal to any settlement or agricultural activity in the area throughout Early Christian, Viking and Medieval times.

7.3.5.4 Zone DT-3. This zone opens when clearance of the remaining woodland is suggested by falling curves for all tree species and *Ilex* disappears completely. *Pinus* reappears at 6 cm and must reflect some of the local plantations of the nineteenth/twentieth centuries. Cultivation of *Hordeum* type is evident throughout the zone and the high values of Gramineae, *Plantago lanceolata*, Leguminosae and Compositae tub./lig. at 10 cm suggest emphasis

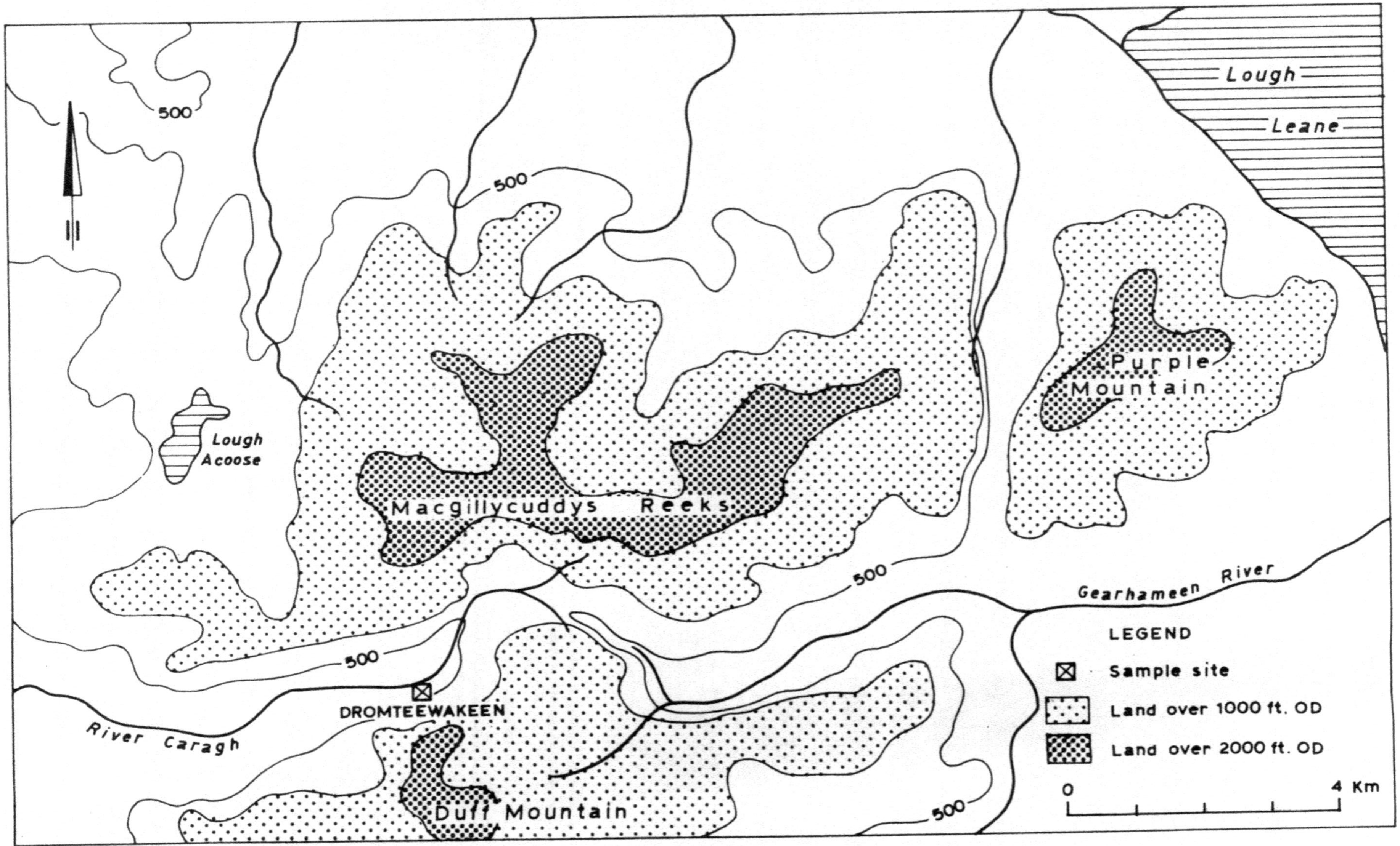

Fig. 12 Location of Dromteewakeen sample site.

on pastoralism. The local peat vegetation has changed little with the exception of high Cyperaceae values above 10 cm and a slight rise in the Potentilla t. curve which may be due to increased grazing pressure on the peatland.

It is tempting to correlate this phase of intensive activity on the peat-covered Dromatouk hillside, with that of Cashelkeelty's Zone CK-6, which has been attributed to the eighteenth/early nineteenth century pressure on land brought about by rapid population growth (see p. 97 above). The post-Famine decline is seen in the falling weed taxa curves between 5-10 cm.

7.3.6 Summary of major events at Dromatouk I

(I) Charcoal dated to 1380 ± 50 bc may possibly relate to the construction/ use of the nearby stone alignment.

(II) Podzolization and gleying were taking place for c 1000 years before 70 bc, under conditions of spreading heathland, some open woodland and little or no human activity.

(III) Formation of blanket peat began on the site at c 70 ± 30 bc.

(IV) Shortly after 70 bc, some pastoral farming was practised on the hillside.

(V) From Early Christian to Medieval times, human activity in the area was limited to one phase of pastoral farming.

(VI) A period of mixed farming on the hillside is attributed to the pre-Famine (1840s) years when woodland cover is negligible.

(VII) Farming activities decline—probably as a direct consequence of the Famine.

7.4 Dromteewakeen I, Co. Kerry (Fig. 21, Table 12).

7.4.1 Site Description

Townland: Dromteewakeen Altitude 101 m OD
County: Kerry Nat. Grid. Ref. V760 081

A monolith of peat and mineral soil was taken from the vicinity of Dromteewakeen stone alignment. The siting of this alignment is rather unusual in that it is located on the floor of an upland valley, in the midst of the Macgillycuddy Reeks. This narrow valley (averaging 400 m wide) is occupied by the westward-flowing Caragh river which eventually enters Dingle Bay, and in the region of Dromteewakeen, the valley is virtually enclosed by the towering Reeks which average 800 m in altitude (Fig. 12).

The lower mountain slopes support extensive tracts of blanket peat which have spread onto the valley floor right to the river's edge. Some drainage of the valley peat has been carried out in the past few decades and the alignment is now virtually surrounded by meadows and rough pasture which have been reclaimed from the peat. The Caragh valley is sparsely populated today and the few small holdings which do exist have economies based mainly on sheep and cattle rearing supplemented by a little root crop cultivation. The local vegetation is dominated by that of the blanket peat and with the exception of isolated examples of Ilex and Alnus by the streamsides, the only tree cover is that planted in the vicinity of the homesteads.

The character of the Dromteewakeen pollen diagram is most certainly local or extra-local with the pollen catchment area being restricted to that part of the Caragh river valley which runs east-west through the Macgillycuddy Reeks.

7.4.2 Sampling and Stratigraphy. A pit (c $1\ m^3$) was dug c 5 m north of the stone alignment and a monolith taken. The stratigraphy was as follows:

1-31 cm	Dark brown fibrous peat with considerable root penetration.
32-41 cm	Dark brown peat with a little root penetration.
42-46 cm	Black peat with much charcoal.
47-56 cm	Grey gleyed, stony boulder clay.
> 57 cm	Orange illuvial horizon.

7.4.3 Presentation of Data. All pollen values were calculated as percentages of total pollen excluding spores. Sampling intervals were generally 5 cm and the results are presented in Fig. 21 and Table 12. For discussion purposes, the peat spectra are divided into local pollen assemblage zones.

7.4.4 Radiocarbon Dating. The basal 2 cm of peat (having removed all visible traces of charcoal) were processed at the Groningen laboratory:

GrN-9283: 985 ± 45 bp or 965 ad.

Slight contamination of the sample with flecks of charcoal is a possibility and the date may therefore be slightly too young.

7.4.5 Vegetational History

7.4.5.1 The Pre-Peat Vegetation. Pollen was only preserved in the gleyed (G) horizon and the three spectra from this layer shall be discussed together in order to give an indication of local conditions prior to peat formation.

In all three spectra, Corylus is the dominant woodland species and together with Betula, Alnus and some Quercus and Ilex it may have formed pockets of woodland in the valley and on the hillsides, or alternatively dispersed trees around the site may be represented (average AP% = 31.3%). It is likely that much of the Alnus and the Salix were growing along the banks of the river Caragh. The high Ericaceae values reflect local heathland conditions and pockets of peat may even have been present in the locality as evidenced by the pollen of Potentilla t., Myrica, Cyperaceae, Sphagnum and Narthecium.

In spite of the increasing acidification of the soil however, considerable agricultural activity was taking place in the valley prior to 965 ad. Mixed farming is suggested by the cereal grains of Triticum t. and Hordeum t. together with the pastoral indicators Plantago lanceolata, Ranunculus, Leguminosae and Compositae tub./lig. The dating or context of this activity is difficult to resolve. Since the pollen preserved, is only that deposited since acidification of the soil reached a sufficiently high level (i.e. pH > 5.5), the farming activity just outlined must have been taking place during the podzolization period immediately before peat initiation. The time-span of this podzolization and gleying period is however, unknown and even though it has been suggested (see p. 101 above) that at Dromatouk, c 1000 years was involved, this can vary enormously, depending on local topographical details.

One may simply conclude therefore, that considerable farming activity was taking place at Dromteewakeen for some time before 965 ad possibly during Early Christian and Viking times. In view of the nearby Early Christian monastic settlement at Inisfallen (c 13 km to the northeast) activity of that period is not improbable in the Dromteewakeen area. It is probable that this farming was confined to the valley itself, while heathland conditions and peat growth were spreading on the surrounding hillsides.

It is not possible to correlate the period of construction/use of the nearby stone alignment with any stage of vegetational history as outlined in the pollen diagram. On the basis of other sites examined, a date of c 1300-1000 bc is being suggested for the stone alignments in general and there is nothing to suggest that Dromteewakeen was any later. It is therefore suggested, that during the construction/use of the alignment, soil conditions were such that breakdown of the humus layer with consequent leaching and podzolization had not yet begun and the low level of soil-acidity did not allow for pollen preservation. On the other hand, there is the remote possibility that podzolization was in progress for c 2000 years and that pollen from the alignment period is represented in the spectra just described, but this cannot be proved.

7.4.5.2 Zone DW-1. This zone is characterised by relatively high *Corylus* values together with *Betula*, *Quercus* and *Alnus* and constant but low values for *Salix* and *Ilex*. Little change is evident since the pre-peat period. Some mixed farming is suggested by the two *Triticum* type grains, the appearance of *Rumex* a.t. at 38 cm and the continuous curves for *Plantago lanceolata*, *Ranunculus* and Compositae lig. The local peat vegetation was dominated by *Myrica* with Ericaceae and Cyperaceae also present.

The humification level of the entire peat profile was fairly uniform with just one gradual change from medium to low levels at c 31 cm, which suggests a fairly constant rate of formation. This rate of peat formation has been calculated at approximately 1 cm per 23 years and based on this, a rough interpolation of dates is attempted for the zone boundaries. It should be stressed however, that these dates should not be considered as anything more than the roughest approximations.

It is likely that Zone DW-1 covers the eleventh and twelfth centuries (965-c 1218 ad), a period when Ireland was divided up into one to two hundred kingdoms between which there was constant rivalry and warring (O Cuiv, 1967, 116). In 1169, the Norman invasion marked the beginning of widespread change and development, but even though the northern part of Co. Kerry was overrun by the Fitz Thomas branch of the Geraldine family, the remainder of the southwestern corner of Ireland remained under the sway of the Gaelic McCarthys and O'Sullivans (Martin,1967, 141). One may therefore assume that what little activity was taking place in Dromteewakeen at this time was that of the local Gaelic farmers.

7.4.5.3 Zone DW-2. The opening of this zone is marked by clearance of *Corylus*, *Betula* and *Quercus*, corresponding with a rising Gramineae curve, together with dramatic peaks of *Rumex* a.t., *Cerealia* and *Potentilla* t. The latter peak (25.8%) may not in this instance, be related to grazing pressure since the curves of other pastoral indicators such as *Plantago lanceolata* and

Ranunculus remain constant at this level (33 cm), but the high values may have resulted from Potentilla t. growing at the sample site as a component of the peat vegetation. Local wet conditions are indicated by the rising values of Cyperaceae.

The occurrence of 15 grains of Fagus and a single Juglans (walnut) grain at the lower zone boundary (33 cm) is worthy of comment. It has been generally accepted (Jessen, 1949; Mitchell, 1965, 131; Smith, 1970, 82) that Fagus did not reach Ireland in early post-glacial times (probably due to a slow migration rate) but first appeared in the country in the seventeenth century when it was a favoured species for plantation. As a result, the many isolated findings of Fagus pollen in earlier phases of pollen diagrams, have been attributed to long distance transport from southern England where the beech was established possibly as early as Neolithic times (Godwin, 1975, 276).

Bearing in mind the probably inaccurate dating, the Dromteewakeen Fagus grains are likely to date to somewhere in the thirteenth century. The explanation of long distance transport of the grains by wind, is not considered plausible in this particular instance for the following reasons:

(I) The restricted pollen catchment area of the site due to its extremely sheltered position.

(II) The unsuitability of Fagus pollen grains for wind transport. The pollen productivity of Fagus flowers is low, the grains are large and heavy and have a tendency to sink directly to the forest floor and so escape entry into ascending air currents [illegible] 1975, 273).

(III) The finding of 15 grains in one sample, in view of point (II) above, is hardly due to coincidence.

Fagus grains have also been identified in the profiles from Cashelkeelty I, Cashelkeelty III and Cashelkeelty IV but all may be attributed to the historic plantation period.

It is therefore tentatively suggested here, that the Fagus pollen record at Dromteewakeen reflects the presence of that tree in the locality in the thirteenth century. Human activity has already been established for the area during this period and one could perhaps speculate as to the probability of a Norman introduction of exotic trees, including Fagus and Juglans (the question of the presence of Juglans in Ireland shall be discussed in detail in section 7.6 below). The problem remains however, of finding a suitably dry calcareous habitat for Fagus, in the Dromteewakeen area. The suggestion of a Norman introduction of Fagus is of course merely speculation, closer sampling of deposits from the period shall have to be undertaken and accurate dating methods applied.

This phase with its emphasis on arable farming seems to have lasted for the greater part of the thirteenth century before the cleared areas were abandoned and allowed to revert to scrubland consisting mainly of Corylus and Betula. By the beginning of the fourteenth century however, there is evidence of renewed clearance of the scrub and a phase of mixed farming is indicated until the end of the eighteenth century.

It is impossible to correlate these different Medieval phases with precise historical events or periods without a detailed study of the local history which was considered to be outside the scope of the present work.

7.4.5.4 Zone DW-3. This is a zone of continuous farming activity covering the period from the end of the seventeenth century until recent times. Both pastoral and arable farming were carried out and the cereals cultivated included *Triticum* t., *Hordeum* t. and *Avena* t. There is no evidence of the pre-Famine intensification of land use as was seen at Cashelkeelty and Dromatouk, and in fact the Dromteewakeen valley appears to have been constantly occupied, with varying degrees of intensity, from early historic times right through to the present day.

7.4.6 Summary of major events at Dromteewakeen

(I) Construction of stone alignment c 1300-1000 bc

(II) Earliest preservation of pollen indicates a mixed farming economy, possibly during Early Christian times.

(III) Initiation of peat formation at c 965 ad.

(IV) Continuation of apparently small-scale mixed farming, during the eleventh and twelfth centuries.

(V) Intensification of farming activities in the thirteenth century, with emphasis on arable farming and possible introduction of *Fagus*.

(VI) After a brief lull in activities at the end of the thirteenth century, both cereal cultivation and pastoralism were carried out throughout the entire Medieval period and continued to the present day.

7.5 Maughanasilly I, Co. Cork (Fig. 22, Table 13)

7.5.1 Site Description

Townland: Maughanasilly	Altitude:	213 mOD
County: Cork	Nat. Grid Ref.	W105 058

A monolith of peat and mineral soil was removed during excavation, from the sectional profile which ran east-west along the centre-line of the alignment. As already mentioned (see p. 69 above) this alignment is sited in an upland valley occupied by a tributary of the Coomhola river. The monument itself is located on the crest of a small boulder-clay hillock on the northern slopes of Knockbreteen mountain (243 m OD) with Lough Altooreeen lying c 300 m to the southwest. This valley which is c 2.5 km long is hemmed in on the north, east and south by mountains of the Shehy range and a narrow gap to the west gives access to the Coomhola valley (Fig. 9). The pollen catchment area represented in the Maughanasilly diagram is therefore likely to be limited to this enclosed valley.

The landscape of the Maughnasilly valley is today dominated by vast stretches of blanket peat which supports *Molinia caerulea*, *Erica tetralix* (cross-leaved heath), *Calluna vulgaris* (ling) and *Scirpus tabernaemontani* (glaucous club-rush). *Myrica gale* is abundant in the area but noticeably confined to the drier roadsides. The trees are similarly confined to the roadsides and include Ilex, Salix, Alnus, while along the banks of streams, Alnus, Betula and Salix are common. The valley today supports one farmstead, the economy of which is based primarily on sheep (which graze freely on the hillside) with a few milch cows and small-scale cultivation.

7.5.2 Sampling and Stratigraphy. The monolith was removed in metal boxes, from a point 1.8 m east of the easternmost stone of the alignment. The stratigraphy was as follows:

1-10 cm	Brown, dry fibrous peat.
11-23 cm	Black well-humified peat with charcoal in its basal layer.
24-30 cm	Grey, stony gleyed boulder clay.
31-32 cm	Black humic layer — A_R horizon
33-59 cm	Orange/yellow illuvial horizon with root channels.

7.5.3 Presentation of Data. All pollen values are expressed as percentages of total pollen excluding spores. Results are presented in Fig. 22 and Table 13. Local pollen assemblage zones have been established for discussion purposes.

7.5.4 Radiocarbon Dating. Two radiocarbon dates were obtained from samples collected during excavation:

Table 6: Radiocarbon dates relating to Maughanasilly I

Laboratory Number	Context	C^{14} Date bp	C^{14} Date bc
GrN-9280	Charcoal from top of gley (G) horizon	3265 ± 55	1315
GrN-9281	Charcoal embedded in the basal peat	3265 ± 55	1315

Both radiocarbon dates were measured at the Groningen laboratory, and received acid pretreatment.

7.5.5 Vegetational History

7.5.5.1 The Pre-Peat Vegetation. The pollen preserved in the G and A_R horizons represents that deposited after the onset of leaching, which resulted in an increase in soil acidity. The construction of the stone alignment took place before the final gleying and for reasons already outlined (see p. 69 above) the upper cm or top of the G horizon, is considered to be the 'activity level' associated with the construction/use of the momument.

The general picture of the immediate pre-peat period is one of quite open woodland comprising Corylus, Betula, Alnus, Ilex and Quercus. Pinus was present, probably on the shallower soils of the hillsides and at one stage, actually grew on the boulder-clay hillock itself, as evidenced by the root channels which contained Pinus sylvestris charcoal. This charcoal was dated to 3265 ± 66 bc or 1315 bc, and it is probable that the burning of the Pinus trees on the hillock accounts for the sudden disappearance of Pinus pollen in the upper two samples (24 and 25 cm) of the G horizon.

Heathland must have been widespread on the hillsides as indicated by the high Ericaceae values and even though the extremely high values of Polypodium and Dryopteris spores must in part be due to differential preservation, both ferns must have been abundant in the area. The rather low Gramineae values point to limited open grassland but the presence of Plantago lanceolata, Ranunculus, Leguminosae and Compositae tub. and lig. suggests some pastoral activity.

It has been argued above (p. 71) that the date of 1315 bc obtained for the Pinus clearance, is likely to relate to the construction of the alignment. If this is indeed the case, then in middle Bronze Age times when the alignment builders (and possibly also the nearby circle builders) were living in the valley, soils were degenerating and farming activity was restricted to pastoralism. It is possible of course, that since the pollen record only refers to the period of podzolization, we are simply picking up here the tail-end of Bronze Age activity in the valley, when lands previously used for cultivation were either lying fallow or being used as pasturage, before final abandonment of the area.

7.5.5.2 Zone MS-1. The date (1315 bc) obtained for Pinus charcoal embedded in the basal peat, is identical to that obtained from the upper levels of the G horizon which suggests that a short period of time elapsed between the burning of the Pinus trees and the spread of peat onto the hillock.

This zone is characterized by increased woodland cover (maximum total tree pollen of 70%) dominated by Ilex and Corylus together with Betula and Quercus. The sudden, sharp peak of Ilex at 22 cm corresponds with similar peaks in the curves of Corylus and Betula and seems best interpreted as invasion of abandoned clearings. The Gramineae curve drops at this point and the weed taxa disappear. This abandonment of the Maughanasilly area may have lasted throughout late Bronze Age and Iron Age times, but interpolation of dates was not attempted because a change in the peat formation rate is indicated by changing humification at 10 cm.

7.5.5.3 Zone MS-2. The opening of this zone sees a drastic decline in the woodland cover (total tree pollen = c 20%) to values similar to those of today. This clearance which may have begun in Early Christian times, was accompanied by rising Gramineae values, the reappearance of Plantago lanceolata, Leguminosae and Ranunculus and the occurrence of cereal type pollen grains. Mixed farming was carried out and the two peaks of the Potentilla t. curve at 15 cm and 5 cm may be the result of grazing pressure. Cereals cultivated included Hordeum t., Triticum t. and Zea mays (maize). The single characteristic grain of Zea mays is of interest because maize was not introduced into Ireland until the 1840s (as famine relief) after which time some experimentation with its cultivation took place but never really succeeded. Today, what little maize is grown in S Ireland is used as a fodder crop.

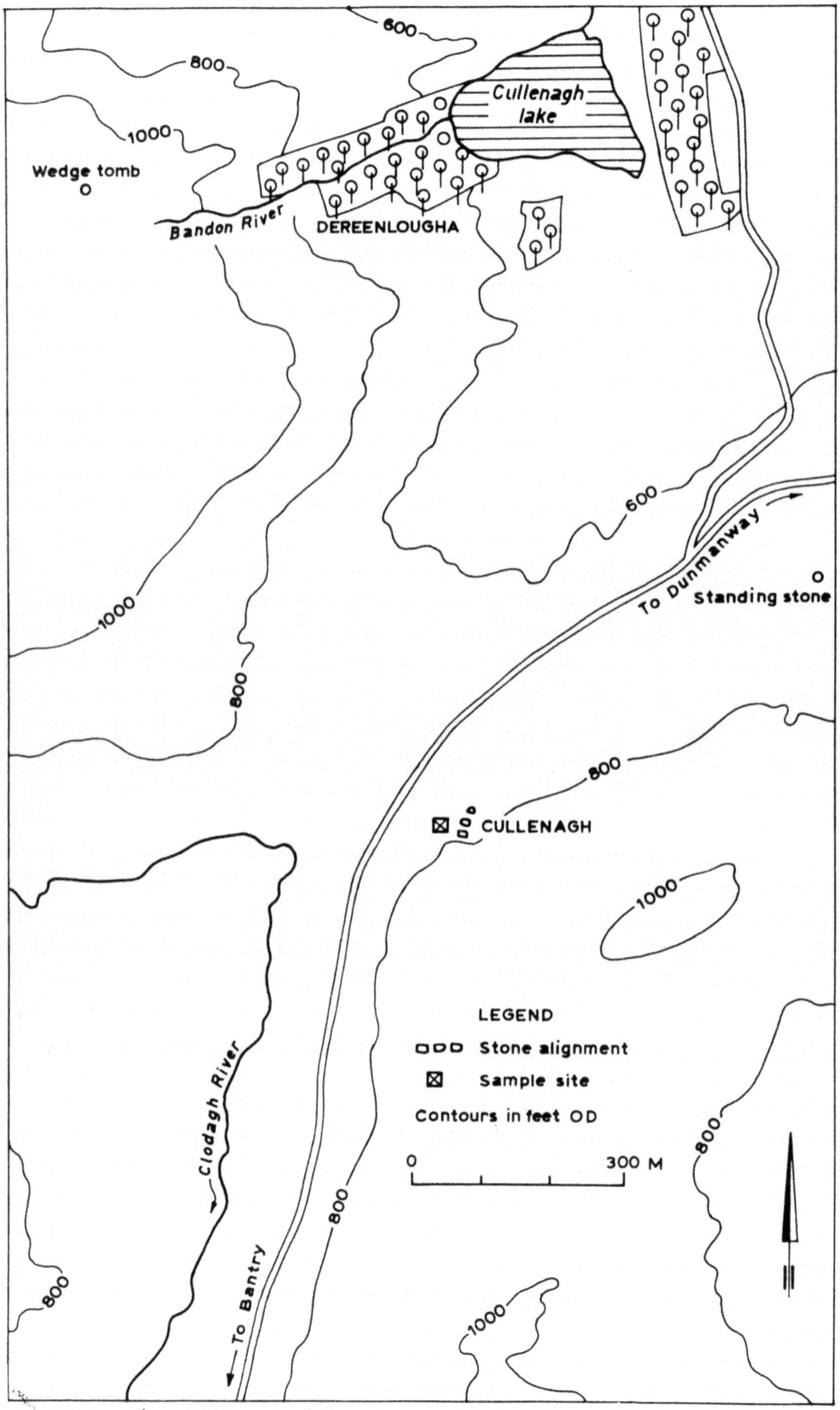

Fig.13 Location of Cullenagh alignment and sample site.

The reappearance of Pinus at 5 cm may be attributed to the historic plantations.

7.5.6 Summary of major events at Maughanasilly.

(I) Spreading heathland and some open woodland with limited pastoral activity, prior to the fourteenth century bc.

(II) Burning of Pinus trees on the hillock around 1315 bc, possibly in connection with construction of the stone alignment.

(III) Initiation of peat growth on the hillock shortly after 1315 bc.

(IV) Area is abandoned and woodland regeneration takes place.

(V) Clearance of secondary woodland occurs, possibly in Early Christian times and continuous mixed farming is carried out in the valley up to the present day.

7.6 Cullenagh I, Co. Cork (Fig. 23, Table 14)

7.6.1 Site Description	Townland:	Cullenagh	Altitude:	244 mOD
	County:	Cork	Nat. Grid Ref.	W116 053

This monolith was taken from a peat deposit in the vicinity of Cullenagh stone alignment. The sampling site is located on a level ledge on the northwestern slopes of a steep ridge (325 mOD) in the Maughanaclea Hills (Fig. 13). To the southwest, the site overlooks the enclosed valley of the westward-flowing Clodagh river while to the northeast, more open-type terrain is present which contains Cullenagh Lake and the headwaters of the Bandon river which flows eastward. Therefore, even though the sample site is sheltered from the prevailing southwesterly winds, its pollen catchment area is likely to include the more open Bandon valley to the northeast from which some regional pollen may be expected to derive.

The hillsides and to a certain extent, the lower valley floor, are now covered by blanket peat with its characteristic vegetation as already described for Cashelkeelty etc. The valley supports about three small farmsteads and extensive drainage has been carried out on the lower hill slopes to provide small pockets of land suitable for pasturage and a little cultivation. At a distance of c 2 km to the east there are extensive woodlands planted on the Coolkellure House estate.

7.6.2 Sampling and Stratigraphy. A pit (c 1 m^3) was dug c 5 m north of the stone alignment on the level ledge, and the peat was removed from the exposed face in a metal box. The mineral soil samples were collected separately in glass tubes. The following is an outline of the stratigraphy:

1-22 cm	Brown peat of intermediate humification with visible Eriophorum (cotton-grass) remains.
23-35 cm	Highly humified black, greasy peat.
36-42 cm	Grey, gleyed stony boulder clay.
c 42 cm	Iron Pan.
43 cm	Orange, illuvial horizon.

7.6.3 Presentation of Data. The pollen values have been calculated as percentages of the total pollen, excluding spores. The results are presented in Fig. 23 and Table 14. Local pollen assemblage zones have been established for discussion purposes.

7.6.4 Radiocarbon Dating. The basal 2 cm of peat (whole fraction) were radiocarbon dated at the Groningen laboratory:

GrN-9282 975 ± 45 bp or 975 ad.

7.6.5 Vegetational History

7.6.6.1 The Pre-Peat Vegetation. Pollen preservation in the mineral soil was extremely poor and only the sample from the top of the gleyed (G) horizon (36 cm) provided a sufficiently high pollen sum for reliable interpretation.

The tree-pollen represents 30% of the total, which suggests that either woodland was present within 100 m of the site or alternatively, the site was surrounded by dispersed trees (Tinsley and Smith, 1974, 563). *Betula* and *Corylus* were the dominant species together with a little *Alnus* and *Quercus*. Open grassland must have been present as indicated by the Gramineae values (15%) and both arable and pastoral indicators suggest considerable human activity. The cereal type pollen grains were too corroded for identification to type level. Heathland must also have been present in the locality.

The radiocarbon date of 975 ad provides a *terminus ante quem* for the activity just outlined and considering the poor preservation of pollen in the lower soil samples, the time lapse between deposition of the pollen at 36 cm and peat growth cannot have been too long. It may well be, therefore that this spectrum relates to Early Christian/Viking times. As at Dromteewakeen (see p. 105 above) the stone alignment phase of activity cannot be recognized in in the diagram. Considering the late date of peat formation at the Cullenagh site, it is probable that the podzolization process (and by implication, the preservation of pollen) had not started at the time of construction or use of the monument.

7.6.5.2 Zone Cu-1. This zone is characterized by almost complete disappearance of woodland, and the presence of extensive grassland with pastoral and arable indicators. All tree species have been reduced to negligible levels and the few *Pinus* grains are probably the result of long distance transport by wind. The Gramineae values are high (c 30%) at 35 and 30 cm but fall to 7% at the upper zone boundary. Pastoral farming must have been practised in the locality and the peak of *Plantago lanceolata* at 30 cm is matched by similar peak in the curve of *Potentilla* t. The cereal being cultivated was *Triticum* t. The local peat vegetation was dominated by Ericaceae, Cyperaceae and *Potentilla* t.

The duration of this zone is not known but it is likely that several centuries after 975 ad are represented which would cover the period of Norman invasion and occupation.

The identification of five pollen grains of *Juglans* at 30 cm (Table 14) suggests that this tree was growing in the locality, possibly during Norman times.

The *Juglans* pollen grain is large, oblate spheroidal in shape and not very suitable for long distance transport by wind. The history of this tree in Britain and W Europe appears to have been closely related to the spread of Roman influence and the isolated grains from the peat bogs of Galloway and Devon are considered by Godwin (1975, 248) to be 'certainly due to long-distance transport'. The appearance of *Juglans* at Old Buckenham Mere however, is at the Norman level (Godwin, 1975, 248).

On the basis of the *Juglans* grains found at Cullenagh I (an isolated grain also occurs at 15 cm) and the single grain identified in a probable Norman context at Dromteewakeen (see p. 106 above), it is postulated that this exotic species, together with *Fagus*, was introduced by the Normans once occupation had been established and their attention had turned to agricultural organization and estate management. In order to test this hypothesis however, more detailed pollen analyses shall have to be undertaken and the history of Norman occupation/activity in the area, closely examined. The results should be worthwhile however, and could provide details regarding the introduction of these exotics to Ireland.

7.6.5.3 Zone Cu-2. This zone which must cover the greater part of the Medieval period, is characterized by a slight regeneration of woodland, especially *Corylus*, *Betula* and *Quercus* and continued farming activity, both pastoral and arable. The peaks in the Gramineae curve at 5 to 10 cm may in part be due to local bogland species (e.g. *Molinia caerulea*). *Narthecium ossifragum* has also appeared on the bog surface.

A phase with emphasis on pastoral farming is indicated at 20 cm by the high peaks in the curves of *Plantago lanceolata*, *Ranunculus* and *Potentilla* t. These curves have fallen quite sharply by 15 cm but cereals (*Triticum* t. and *Hordeum* t.) continue to be cultivated.

It seems reasonable to conclude that the Cullenagh area was continuously occupied since at least Early Christian times.

7.6.5.4 Zone Cu-3. The surface sample representing modern conditions, shows a rise in all tree pollen values which must reflect the local plantations and a decline in human activity which now seems restricted to pastoral farming.

7.6.6 Summary of major events at Cullenagh I

(I) Construction of the stone alignment at an unknown date.

(II) Pastoral and arable farming take place before 975 ad — possibly during Early Christian or Viking times.

(III) Peat growth starts around 975 ad.

(IV) During (?) Norman times the woodland is much reduced, pastoral and arable farming is carried out and *Juglans* is probably introduced to the area.

(V) Throughout the Medieval period there is continued farming activity.

(VI) Slight woodland regeneration/planting and a decline in human activity in recent times.

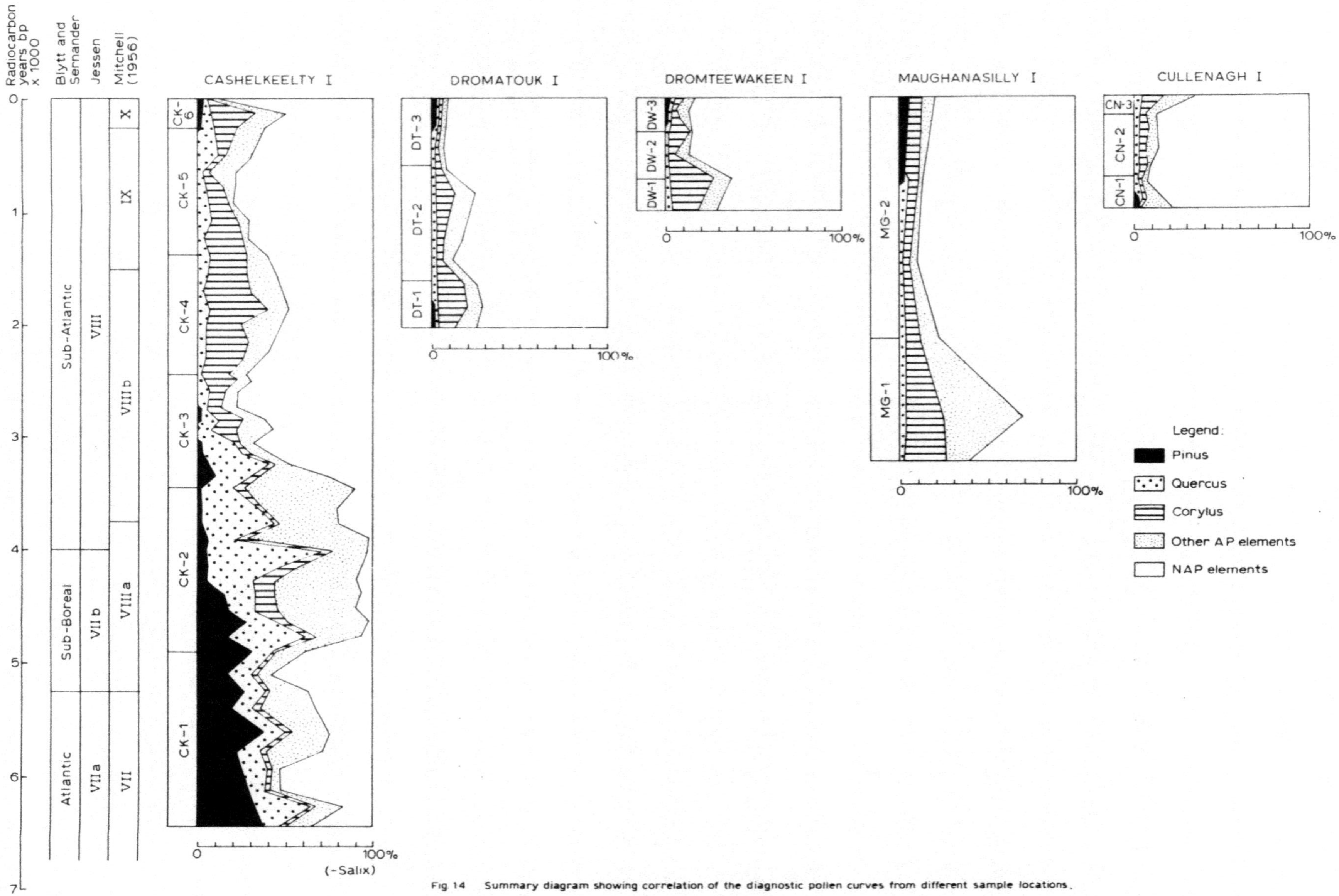

Fig. 14 Summary diagram showing correlation of the diagnostic pollen curves from different sample locations.

7.7 General Summary

The information presented in the preceeding sections has been summarised in Fig. 14, in which the more diagnostic pollen curves from each sample location are plotted against the radiocarbon time-scale. This should enable the reader to see at a glance, the general trends through time of woodland development, clearance and regeneration, in the different localities examined.

It seems clear that the disappearance of *Pinus* after the first millennium bc, was a widespread phenomenon throughout the study area and that *Quercus* had, in general, become greatly reduced by the mid-second millennium bc. The fluctuations in the *Corylus* curve and AP/NAP values, vary from one site to another, depending on the intensity and nature of local human activity.

CHAPTER 8

DISCUSSION OF RESULTS
AND THEIR ARCHAEOLOGICAL IMPLICATION

8.1 Introduction

At the outset, this study was intended to investigate the environmental setting and subsistence economy of a single group of people, the stone alignment builders, of SW Ireland. It has provided however, a general scheme of environmental (in particular vegetational) development and change throughout the prehistoric period in the study area, integrated as far as possible with what little evidence the archaeological record has provided regarding economic and cultural systems. The main reason for this change in emphasis is lack of basic data concerning the group in question — the nature and location of settlement sites in unknown, details of material culture have, at best, remained vague and the chronological context is uncertain. The two excavations carried out as part of this study, emphasized the difficulties in retrieval of the necessary data rather than solving any of the problems. Large-scale, intensive survey and excavation is needed before anything approaching a synthesis of the Bronze Age sociocultural systems of SW Ireland can be attempted.

This study may therefore be considered as the first tentative step towards the final goal. The preliminary and speculative nature of many of the results cannot be over-stressed. Much of the vegetational history of the prehistoric period is inferred from a single profile (Cashelkeelty I) in a single bog and many local variations (particularly in terms of human interference) must have existed throughout the study area. In the absence of supporting archaeological evidence, inference of specific agricultural activities from the palynological record is fraught with difficulties. Certain imprecisions are inherent in the radiocarbon chronology and calibration will introduce additional changes. Such a litany of gloom does not however, negate the results outlined in the preceding chapters. The model presented of prehistoric developments in SW Ireland is but one of the possible explanations of the observed empirical palaeoecological and archaeological data and as such, should serve as a stimulating basis for discussion and provide possible channels along which future research could be most fruitfully directed.

The following discussion is organized under the traditional chronological headings Neolithic, Bronze Age, Iron Age and Early Christian period. It should be remembered that the episodes deduced from palynological evidence are assigned to earlier Neolithic, later Neolithic etc. by virtue of their radiocarbon dates and these rather arbitrary divisions of radiocarbon time are merely intended to facilitate discussion of chronology in general terms.

8.2 The Earlier Neolithic

The recording of human activity, including cereal cultivation, in the Cashelkeelty area of Co. Kerry between 3895 and 2965 bc has far-reaching implications for the beginnings of the Irish Neolithic. The nature of this early Cashelkeelty agriculture is not clear but it is suggested that clearances in the woodland were carried out using the slash-and-burn technique, followed immediately by cereal cultivation (both wheat and barley) which in time appears to have given way to pasturage before the clearings were finally abandoned and woodland allowed to regenerate. The extent and duration of these early Neolithic clearances cannot be deduced from present evidence and neither do we have any indications of the material culture belonging to these early farmers, but their mere presence does at least permit a preliminary evaluation of the introduction of agriculture to SW Ireland.

Perhaps the first implication is that the elm decline (centering on 3000 bc) can no longer be considered as marking the beginning of Neolithic activity. This conclusion is hardly surprising when one considers the numerous 'pre-elm decline clearances' recorded during palaeobotanical studies throughout Britain and N Ireland in the past two decades. Several sites in west Cumberland (Walker, 1966) and around Morecambe Bay (Oldfield, 1963) have indicated disturbance of the woodland cover before the classical elm decline, which according to Pennington (1970, 68) implies " a wide-ranging nomadic economy". At Hockham and Seamere in E Anglia, Sims (1973) has recorded small localised clearing prior to the elm decline which he tentatively explains as the result of incipient agriculture. The first forest clearance at Broome Heath, also in E Anglia, took place c 3500 bc and has been attributed by Dimbleby and Evans (in Wainwright, 1972) to Mesolithic opening of the upland forest for hunting. Similar Mesolithic activities have been postulated to explain early clearances on the North Yorkshire Moors (Spratt and Simmons, 1976) and Dartmoor (Simmons, 1964). This emphasis on a Mesolithic explanation for the British clearances is due in part to the absence of reliable arable indicators such as cereals, and, probably more so, to respect for the prevailing Neolithic chronology. The Artemisia pollen recorded at Storrs Moss by Oldfield can hardly be considered evidence for arable farming.

The Irish evidence for early Neolithic clearances is more convincing. At Ballynagilly, Co. Tyrone radiocarbon dates ranging from 3795 ± 90 bc (UB-305) to 3550 ± 85 bc (UB-559) have been obtained from Neolithic pits containing Lyles Hill pottery (ApSimon, 1976, 20) and Pilcher and Smith (1979, 367) have noted some features of similar date in the pollen record from the site which could be interpreted as anthropogenic. This activity has been termed "Earliest Neolithic" (E'N) by ApSimon (1976) and even though the features span an area 100 m across, the amount of clearance around them, if any, is unknown. A detailed pollen diagram from a raised bog at Ballyscullion Co. Antrim has shown clear indications of human interference with the forest cover between 3865 ± 90 bc (UB-296) and 3580 ± 60 bc (UB-116) which Smith (1975, 64) compares with the Earliest Neolithic phase at Ballynagilly, but the possibility of Mesolithic land use is not ruled out. Nearby at Newferry, cereal pollen was identified in a layer containing Neolithic artefacts and dated to 3330 ± 170 bc (D-36). The beginning of forest clearance and a landnam

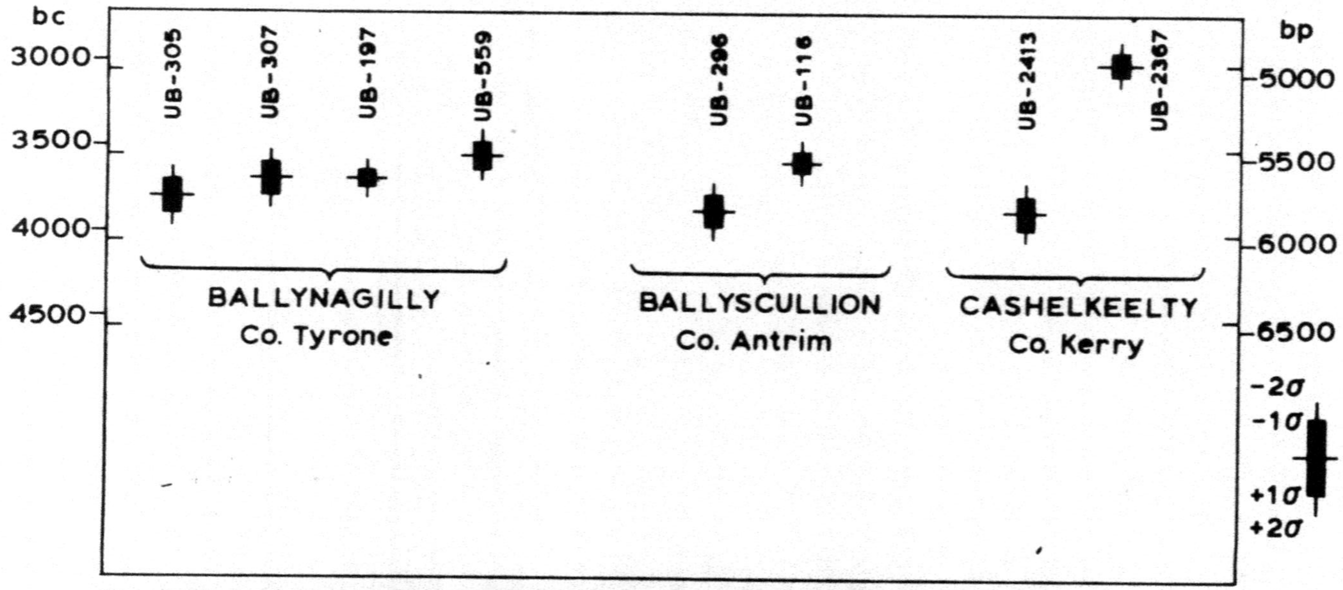

Fig. 15 Plot of earliest Irish Neolithic dates - after Smith 1975 with additions.

phase has been dated to 3380 ± 120 bc (Q-555) at Fallahogy, Co. Derry. Recent work by Teunissen (1980) at Dolan, Connemara has placed cereal pollen in the later part of the Atlantic period but unfortunately, no radiocarbon dates have been obtained.

When the earliest Neolithic radiocarbon dates from Ballynagilly, Ballyscullion and Cashelkeelty are plotted and compared (Fig. 15), the contemporaneity of onset of clearance is striking and there seems little reason not to envisage the introduction of an agricultural way of life to Ireland in the early fourth millennium bc. Cashelkeelty is however, unusual on two points: it is the only site with evidence for cereal cultivation at this early stage and its southwestern location is far removed from the Mesolithic/early Neolithic sites of N Ireland. This latter point raises interesting questions regarding the mode of introduction of agriculture to Ireland and the extent to which the Mesolithic communities contributed to the transition to a Neolithic way of life.

The actual extent of Mesolithic activity in Ireland is uncertain and as Woodman (1978, 140) points out, the apparent concentration of known sites in the northeastern third of the country may be due more to emphasis of research and collection rather than real distribution. The traditionally held view was that availability of flint for the manufacture of artefacts was responsible for the concentration of Mesolithic activity in NE Ireland but if as Woodman (1978, 142) suggests, food availability was more likely to be the major determinant in site location, then there is little reason to suppose that the southwest was not occupied at this time. For the moment however, there is no evidence to support this hypothesis and the possibility of a Mesolithic contribution to the early farming activity at Cashelkeelty cannot be evaluated. On consideration of the late Mesolithic and early Neolithic material from NE Ireland, both Smith (1976) and Woodman (1976) have concluded that there is little evidence for acculturation of an indigenous population through seaborne contacts, and the arrival of people rather than ideas must be postulated in order to explain many of the economic/technological innovations of the early Neolithic. It therefore seems likely that the early farming episodes at Cashelkeelty may similarly be ascribed to newly-arrived colonists.

We seem therefore to be dealing with a demic mode of Neolithic diffusion and the 'colonization' model of Ammerman and Cavalli-Sforza (1973, 344) seems most applicable to the Irish situation. This implies the intentional settlement by a coherent group of people, usually in a distant land. The catalyst for such movements of people is generally considered to have been changes in demographic pressure which in turn may have been brought about by increased sedentarism.

From which lands then did these early Neolithic colonists arrive and which routeways were followed? The new settlers at Cashelkeelty could have travelled overland from the eastern or northeastern parts of the country or alternatively, the area may have been approached by sea from the west. For the moment, the overland route seems least plausible when one considers the dense climax woodland which covered most of the countryside at this time and there is little evidence of early Neolithic activity in the central and southern regions through which the settlers should have travelled. The coastal location of Cashelkeelty however, supports the idea of arrival by sea and Case (1969a, 179) has argued for the use of skin-boats consisting of ox-hides stretched on a light frame, during early Neolithic times.

The origins of these colonists must, for the moment, remain the subject of speculation in the absence of material attributes and reliable chronologies for comparative purposes. Assuming however, that population pressure/ instability provided the necessary stimulus for immigration, then we should be looking for areas within access and with a well-established Neolithic economy in the early fourth millennium bc, as possible lands of origin. On the basis of chronology, S England can hardly be considered a likely source since the earliest radiocarbon date from a Neolithic context is 3330 ± 150 (BM-138) from Hembury (Whittle, 1977, 246). The Breton area of NW France however seems a more likely source, since megaliths dated to 3850 ± 150 bc (Barnenez G) and 3890 ± 300 (Kercado) bear witness to a well-established Neolithic settlement (L'Helgouach, 1973, 303-219). Once embarked on a voyage of exploration, these Breton crews could have been carried to the shores of SW Ireland by following the southern branch of the main Atlantic sea route which was to be used extensively by seafarers in the following centuries (cf. Bowen, 1970). One should of course, also allow for the possibility of settlers arriving by sea from the northern parts of Ireland itself, and it may well be that the river Shannon and its tributaries were used as routeways to the SW. The incentives to travel such distances in order to settle in SW Ireland may have lain more in the unfavourable situation in the home territory (e.g. overcrowding or political unrest) rather than in any extraordinary advantages offered by the new lands. With the exception of small quantities of pebble flint found along the beaches and in the boulder clay, there are no native flint supplies in SW Ireland which could have attracted the settlers, but on the other hand, it is an area of diverse ecological zones, the mountains, river valleys and coastal areas offering abundant opportunities for the procurement of food supplies and perhaps most important of all, it was apparently an empty land giving endless opportunity for settlement and expansion.

Whatever their origins, the early Cashelkeelty settlers must be considered as a pioneering group of incipient farmers. The difficulties inherent in settling new and distant lands have been outlined by Case (1969a, 180) and several stages of exploration and experimentation with animal transport and seed cultivation must be envisaged before permanent settlement took place. Even then, the going must have been rather difficult while much labour and time was expended in forest clearance and cultivation, before a 'stable adjustment' could be made to a favourable environment. Throughout this pioneering phase, the Cashelkeelty settlers must have exploited the nearby marine and riverine resources in order to supplement their rather precarious diet. According to Waterbolk (1968, 1096) such early Neolithic coastal adaptation would have allowed a high degree of sedentary occupation which in turn, would have made it easier to experiment with agriculture. Hunting and the collection of wild fruits and nuts in the forested areas may also have played a major part in their economy before the farming cycle became established. The major preoccupation of these people must have been survival, and few refinements or luxuries in terms of pottery or elaborate burial monuments are likely to have belonged to such pioneering phases of settlement. It is this very ephemeral nature of the early Neolithic settlements which has apparently militated against their recovery in the archaeological record.

It is of interest to note that the second phase of clearance activity at Cashelkeelty which covers the period from c 3400-c 3000 bc, is contemporary with the beginnings of megalith construction in the Carrowmore region of NW Ireland. Recent excavations have provided radiocarbon dates of 3290 ± 80 bc (Lu-1441) and 3090 ± 60 bc (Lu-1698) for the construction phases of a dolmen (Grave No. 7) and tomb with cruciform chamber (Grave 27) respectively and on the basis of these dates, Burenhult (1980, 115) suggests that the tombs and circles at Carrowmore probably represent the "megalithic introduction" in Ireland. The settlements belonging to these tombs are now the focus of excavation and it will be interesting to see the results concerning the social and economic background of the early Neolithic settlers.

8.3 The Later Neolithic

The fate of the first Cashelkeelty settlers is unknown, and by the beginning of the third millennium bc the area had been deserted and was covered by dense oak woodland. This state of affairs persisted until c 2000 bc and so covers the entire later Neolithic period. It is difficult to know whether this recession was simply a local phenomenon or whether it was widespread over the study area in general. Two possible explanations may be invoked to explain the apparent abandonment of Cashelkeelty: (a) the farmers simply moved to new lands which lay outside the pollen catchment zone of Cashelkeelty and their activities no longer registered on the pollen record, or (b) the group(s) were decimated by some catastrophe.

The former explanation is difficult to accept since there seems little reason to have abandoned an area in which much effort of clearance and management had been invested and no evidence was recovered to suggest climatic or soil degeneration which could have caused such a migration. If the settlers did move, then the new lands occupied throughout the third millennium should be recognized, but with the exception of the late Neolithic settlement on the shores of Lough Gur, Co. Limerick (which lies c 160 km to the northeast), all late Neolithic activity so far identified has been in the eastern and northern parts of the country. It should be remembered that at the time of this apparent recession in the southwest, the floruit of megalith building was reached in the north and east with the construction of court-tombs and passage-tombs (see p. 15 above).

If the Cashelkeelty farmers did not migrate then one must assume some catastrophe, for example a disease epidemic, famine or crop failure which would have been disastrous for communities not yet in 'stable adjustment' with their environment (cf. Case, 1969, 181).

This concept of discontinuous growth throughout the Neolithic has also been recognized in Britain and discussed by Bradley (1978, 105) and Bradley and Hodder (1979, 96). Towards the middle of the third millennium bc, many large-scale agricultural clearances reverted to forest, fewer burial monuments were built and the number of settlements declined sharply. A possible explanation of this recession is that "Neolithic expansion had temporarily outstripped the capacity of the system to support it" (Bradley 1978, 106) and as a result, a more extensive system of land use was adopted which involved far fewer risks but on the other hand could have meant the breakdown of some social units with less demands upon individual groups. Recovery from this disturbance is evident in early Bronze Age times.

Whether a similar model is applicable to the situation in SW Ireland is impossible to say at the moment, without more basic data concerning the Neolithic period in general, throughout the area.

8.4 Early Bronze Age

The beginnings of recovery from the late Neolithic recession are evident during the period c 2000-1500 bc when a little forest clearance was attempted. The reason for this clearance is uncertain and no reliable indicators of cereal cultivation or pastoral activities were identified. Whatever activity was involved was apparently on a small scale or it could be that the slight changes registering on the Cashelkeelty pollen record represent more widespread clearance some distance away, but confirmation of this can only be achieved through additional sampling in the area.

The archaeological evidence for settlement in the study area at this time, relies on the assumption of an early second millennium bc date for the wedge-tombs (see above p. 15). The resemblance of these tombs to the allées couvertes of Brittany which belong to the second half of the third millennium and their general disimilarity to other Irish megaliths, has favoured the idea of their having been introduced from overseas (presumably from Brittany) with little native influence (Herity and Eogan, 1977, 122). Considering the evidence for little or no activity in the SW of Ireland during the later Neolithic, it is tempting to attribute these first signs of recovery in the first half of the second millennium bc to the wedge-tomb builders of the area. Several of these tombs are located on the Beara peninsula with the Ardgroom group lying c 7 km to the west of Cashelkeelty possibly accounting for the disturbances noted in the forest cover.

The economy of the wedge-tomb builders remains obscure but the emphasis on pastoralism as suggested by de Valera and Ó Nualláin (1961, 112) does not seem likely, at least for the southwestern group. When these settlers entered SW Ireland they were faced with a densely wooded landscape with few traces left of the earlier Neolithic clearances, and it is likely that a fairly extensive system of land use was at first adopted, with hunting and fishing supplementing the agricultural produce. The general distribution of wedge-tombs throughout the western half of Ireland suggests that some contact must have existed with other early Bronze Age communities, e.g. the Beaker-using people. Such contacts with well-established communities would have been vital for the new settlers in terms of initial supplies of livestock and cereal seeds and subsequently for trade. While at present there is no evidence to suggest that the wedge-tomb builders exploited the copper ore deposits of the southwest, it is possible that they were a metal-using society (Case, 1966, 167) and that some of the thick-butted flat copper axes found in the study area (see Harbison, 1979, 101) belong to this period. Access to metal tools and implements would have enabled these newcomers to attain a position of 'stable adjustment' far more rapidly than their earlier Neolithic predecessors who found themselves in similar environmental circumstances, and the building of burial monuments was perhaps the first expression of such 'stable adjustment'.

Even though the economy of the wedge-tomb builders remains unknown, a recent archaeozoological study of the late Neolithic/Beaker period settlement at Newgrange which centres on 2000 bc (van Wijngaarden-Bakker, forthcoming) has revealed interesting details regarding the animal husbandry of the period. Emphasis was apparently placed on the breeding of cattle and pigs as the primary food animals and even though cattle and pigs were killed in a proportion of 3:2, four times as much beef was eaten as pork. Fat was evidently of great importance in the diet. It is possible that transhumance was practised, the cattle being brought to the coastal area for summer grazing. Cereal cultivation also formed part of the subsistence economy of this time as evidenced by the Newgrange finds of carbonised grains of Hordeum var. nudum (8) and Triticum dicoccum (1) which gave the radiocarbon date of 1935 ± 35 bc (GrN-6342).

Whether or not the animal husbandry of the wedge-tomb builders was as specialized as that of the Beaker period settlements is difficult to say, since the mountainous southwest must have imposed different limitations to possibilities from those of the rolling countryside of Co. Meath (e.g. sheep rearing may have been more important in the southwest). Until the settlements associated with the wedges are fully investigated however, evidence from contemporary sites can provide useful information regarding the general economic background of the period.

8.5 Middle/Late Bronze Age

The period from c 1500-500 bc was one of continuous intense activity with just one possible lull between c 1050-850 bc, at least in the Cashelkeelty area. Widespread woodland clearance was taking place and in so far as it could be established, the general farming cycle began with cereal cultivation immediately after clearance, followed by increasing emphasis on pastoralism before areas were abandoned and allowed revert to scrub. The method of farming used, including the range of tools and equipment available, is not known but digging sticks, hoes, spades and perhaps a primitive plough, should be envisaged. Organization of the landscape in the form of division of land into fields, first took place during this period, probably in late Bronze Age times when much of the woodland had been cleared and rough grassland or scrub predominated.

The palynological evidence for such activity in the study area is supported by the widespread appearance of burial/ritual monuments such as alignemts and stone circles which have been assigned to the period c. 1400-700 bc (see above p. 74). The many boulder-burials and cairns are also likely to have been built at this time and there is every possibility that wedge-tomb building continued until c 1200 bc (cf. Case 1969b, 22). Copper mining was taking place on Mount Gabriel at c 1500 bc (on the basis of the single available C^{14} date) and on the basis of similarities in adit types and the presence of stone mauls, seven other localities in the study area are likely candidates as Bronze Age mines (Fig. 3). These are Derrycarhoon, Horse Island, Roaringwater Bay, Ballyrisode and Eyries, all in Co. Cork, and Ross Island and Coad

Mountain in Co. Kerry. The mine near Eyries lies just 10 km west of Cashelkeelty. Recent work by Jackson (1979) where he attempts to estimate Bronze Age copper production in SW Ireland, has shown that a great imbalance exists between the estimated total metal production and estimated production of artefacts in the country. If such estimates are correct, then the inescapable conclusion seems to be that Ireland (in particular the southwest) was a major exporter of copper during the Bronze Age (Jackson 1979, 121).

If one assumes that the distribution of burial/ritual monuments and industrial sites reflects in general terms the settlement pattern of the period, then some speculation regarding the possible subsistence economy may be attempted. Perhaps the most striking aspect of their distribution is the clustering around the headwaters of the major rivers and their tributaries and along the coastal strips of the western peninsulas. Even though the majority of burial/ritual monuments themselves are sites on the hillslopes overlooking the valleys, it is probable that as at Cashelkeelty, the main focus of settlement was lower down in the valleys or on the coastal plain.

One aspect of these site locations is the opportunity they offered to draw on economically complementary resources of a number of different environmental zones. This point is well illustrated by considering the location of Cashelkeelty. The settlement(s) belonging to the burial/ritual monuments were probably located on the nearby coastal plain which, if Mitchell's suggestions are correct (see p. 3 above), was slightly wider than at present. While the initial clearances in the oak forest were being established, Bronze Age man had several exploitable food sources near at hand — the sea and nearby Creanshagh River provided easy fishing while the forests not only provided hunting opportunities but must also have supplied wild plants, fruits and nuts.

Once clearances had been made in the lowland forest, conditions were ideal for cereal cultivation — humus-rich soils were exposed (and probably enriched by ashes) and the climate was relatively warm and dry (see p. 5 above). When the clearances extended onto the nearby uplands it is probable that cereals were planted for the few years immediately following the felling and burning but thereafter, the hillslopes were likely to have been used for grazing much as they are today. Transhumance (booleying) may have been practised, as was the tradition until quite recently in Ireland in lowland communities located adjacent to uplands (Evans, 1957).

There is no reason to suppose that, once farming activities had become well established, the hunting and fishing activities ceased. If the settlers were to maintain an extensive economy and minimize risk, then exploitation of the resources of different habitats must have occurred, resulting in the products of mixed farming being supplemented by those of hunting and fishing. The site catchment of Cashelkeelty must have been strongly influenced by local topography, in particular the high mountains to the south and the sea inlet to the north, and therefore a catchment area of 5 km radius as proposed by Higgs and Vita-Finzi (1972) is hardly applicable. Instead one should envisage exploitation for considerable distances along the coastal plain and since much of the travel must have been by boat, the opposite shores of the Ivereagh peninsula may have been at least the subject of exploratory forays. This linear-type sea or river-orientated catchment area is applicable to the majority of alignment and circle sites which tend to be partly hemmed in by mountain ranges but never distant from a river valley.

The general picture which emerges therefore, is one of a diverse economy in which right through to the sixth century bc, new lands were being cleared of woodland and settled. It is difficult not to attribute these developments, at least in part, to the copper mining in the region. It is reasonable to assume that newcomers familiar with the skills and technology of prospecting, mining and metalworking must have arrived in SW Ireland by Middle Bronze Age times. Whether the alignment/circle builders were connected in any way with the mining and metalworking industry it is not possible to say, but their contemporaneity at least suggests that contact existed between the two groups and that the Bronze Age farmers could have played a part in the extensive trade network established by the metalworkers. The development of a well-orgnized society by middle/late Bronze Age times is suggested by the intensive agricultural, industrial and trading activities accompanied by formal burial and ritual monuments.

The paucity of metal finds from the study area which belong to the period c 1500-800 bc suggests that most of the ore mined was exported — both to other parts of Ireland and to Britain and NW Europe. Trade however, implies a two-way movement of goods and when dealing with such a valuable commodity as copper one would expect some evidence of the goods received in return. Could the gold represented in the hoard found at Cappeen be seen as part of the exchange network? The period 800-500 bc saw great developments in metal-working techniques and trading contacts were extended to N and Central Europe and the Mediterranean area (Eogan 1964, 325). Considerable wealth in the study area is indicated by the bronze hoards and numerous artefacts found, while the pollen record shows renewed clearance with barley now being grown in addition to wheat. If, as seems likely, the ard was used during the Bronze Age at Cashelkeelty then it was most probably introduced at this later stage, when lands which had partly reverted to scrub and grassland were again brought into use and the plough would have been necessary to break up a top-soil which had become heavily consolidated by matted grass roots (cf. Boserup, 1965, 24 and Wailes,1972, 158).

Before long however, the signs of pressure become evident. Podzolization of soils in cleared areas has begun, heathland conditions are spreading and in many localised spots, peat has begun to form. The circle and alignment at Cashelkeelty were built on abandoned or depleted farmland and such locations of burial/ritual monuments in Britain have been taken by Bradley (1978, 112) as suggesting competition for land. Cooler and wetter climatic conditions cannot have helped matters and by c 700 bc the basis of subsistence had shifted to pastoral activities. This emphasis on pastoralism continued until c 500 bc at which time a marked recession in activity occurs.

8.6 Iron Age

Only two sites out of the five examined, showed any signs of human activity in the period c 500 bc-c 500 ad. At Cashelkeelty a pastoral phase is postulated at c 100 bc and cereal cultivation took place in the locality around 180 ad, while at Dromatouk, some pastoral activity took place after 70 bc. Otherwise, there seems to have been considerable regeneration of woodland in the absence of agricultural activity. Alternatively, it could perhaps be suggested that a decline in the copper mining and smelting activities resulted in a reduced demand for wood and allowed some woodland regeneration to take place. The reason for such an industrial decline however is difficult to envisage. This picture of general recession is in marked agreement with the results of other palynological studies carried out in central and N Ireland (see p. 36 above) and with the archaeological record as it stands today, all of which has earned the period the title of "the dark age". In marked contrast with later Bronze Age times, the study area now seems not only to have ceased attracting newcomers but the existing population has apparently decreased drastically. The two hillforts and several promontory forts in the area (see Fig. 3) may have been built during this period as may also have been many of the ringforts, but in the absence of excavation, the extent of Iron Age occupation in the area remains unclear.

The earliest ironworking in Ireland is represented by a find of two iron looped and socketed axes from N Ireland which date to some time towards the end of the 7th century bc (Scott, 1979, 195). Ironworking in general however, does not seem to have become fully established until the first millennium ad, in spite of the evident decline in bronzeworking after the sixth century bc. There is no evidence for the communal movement of settlers who could have been responsible for the introduction of iron technology to Ireland, and instead, Scott (1979, 199) envisages a small number of people being involved in this early ironworking, having possibly arrived as "adventurers and traders".

The reasons why such early ironworkers did not apparently settle in the study area, may perhaps be sought in the changes which occurred towards the end of the Bronze Age. It seems difficult to avoid the conclusion that a collapse of the social system established by late Bronze Age times, occurred in the sixth century bc. It could be that widespread podzolization of the soils and to a lesser extent deteriorating climate, undermined the agricultural subsistence economy on which the industrial sector of the community was dependant. Once the basis of subsistence had collapsed, major social disruption must have followed which resulted in the decline of the copper mining industry and its related trade, all of which as far as we can see, was accompanied by a real decline in population. What emerged from this fragmented society was a realignment of social and economic arrangements which involved the setting up of self-sufficient farming units, represented by the ringforts.

It is understandable why the study area no longer attracted settlers in search of agricultural lands — widespread soil degeneration and rapidly spreading peat must have discouraged any such newcomers. It is more difficult however to account for the apparent total decline in the mining industry since copper must have continued to be in demand and considering the slow introduction of ironworking, its value cannot have suffered as a result. Furthermore, one should expect the early ironworkers to have been attracted to an area with a long tradition of metalworking and which did contain some iron ore deposits, consisting mainly of haematites or siderites (Scott, 1979, 197). For the moment however, we must accept that little contact existed between the early ironworkers and the remaining inhabitants of the study area who apparently lived in relative isolation.

8.7 Early Christian Period

The conclusions which may be drawn from the present study concerning Early Christian subsistence economy, add little to the details which have been gleaned from literary sources, as described above in section 2.5. With the exception of Dromatouk and Maughanasilly where chronology is not clear, all sites exhibit renewed clearance of scrub, and intensification of mixed farming activities. The appearance of *Avena* for the first time, during this period at Cashelkeelty is of interest considering its record as the most important cereal cultivated at this time (p. 27).

The general impression gained of the degree of intensification of agriculture during Early Christian times in the study area, is that it was considerably less than the vigorous renewal of agriculture suggested by Mitchell (1976) for central Ireland. This could be due in part to the marginal location of the sites examined in relation to the known ecclesiastical and secular sites of the period, and it seems reasonable to suggest that the emphasis in location of farmsteads was moving from the mountainous areas which were being rapidly enveloped in blanket bog to the areas of more fertile Brown Earth and Brown Podzolic soils in the lowlands to the east and south (cf. Fahy, 1969). To obtain further details of Early Christian agriculture and environmental change, future palaeobotanical studies should be concentrated in these lowland areas and in the immediate vicinity of the ecclesiastical sites.

8.8 Implications of Calibration

The foregoing discussions and conclusions have been based entirely on C^{14} dates expressed in radiocarbon years, and which have therefore been measured on the assumption that the atmospheric C^{14} concentration has remained constant through time. Since the pioneering work of de Vries (1958) however, several radiocarbon laboratories throughout the world have been testing the constancy of natural radiocarbon by measuring the radiocarbon content of samples of known age and more recently, by using dendrochronologically dated tree-rings from North American bristlecone pine and European oaks. The general consensus of opinion is that there have been variations in the atmospheric C^{14} concentration in the past and therefore corrections are necessary to relate the radiocarbon assys of organic samples to a calendar time scale. Here however, agreement amongst the physicists and statisticians ends and as yet, no universally acceptable radiocarbon calibration is available.

Using the bristlecone pine chronology, Suess (1965, 1967) showed major short-term variations or 'wiggles' in the calibration which meant that for certain periods, radiocarbon measurements could not provide unambiguous dates. The validity of these 'wiggles' has been the subject of much debate during the past decade and various calibration curves of varying degrees of smoothness have been proposed by Damon *et al.*, (1972) Ralph *et al.*, (1973) and Clark (1975). Differences between radiocarbon dates for bristlecone pine wood and from historically dated Egyptian material of the same age, have been pointed out by McKerrell (1975) who suggests that corrections based on the bristlecone pine chronology may not be applicable for some periods, to European samples. More recent work on tree-ring chronologies of NW European oaks has shown that for the period 3600-4550 bp the calibration will tend to be smooth rather than 'wiggly' (Pearson et al., 1977) while for the earlier period 4500-5000 bp, secular variation in the atmospheric C^{14} level result in pronounced 'wiggles' which correspond well with those of Suess (de Jong and Mook, 1979).

The confusion caused by this plethora of differing opinions has meant that archaeologists who wish to calibrate their dates are now faced with a range of options concerning the correction curve to be used, none of which have received the unanimous support of physicists or statisticians. In spite of the many unresolved problems and continuing debate on the subject however, many archaeologists have begun to use calibrated dates, often without mentioning the correction method employed, which has succeeded in confusing the issue even more. All radiocarbon determinations used in this study have been expressed in uncorrected radiocarbon years which, until international agreement has been reached on a method of calibration, will facilitate comparison with other uncorrected dates. However, it was considered worthwhile to discuss briefly, the general results of calibration of the C^{14} dates obtained as part of this study, simply in order to outline the general trend and its possible implications.

The corrected dates have been obtained from the tables published recently by Suess (1978) and the results are presented in Table 7. No account has been taken of the precisions involved in the process of calibration. Both the uncalibrated and calibrated dates of the Cashelkeelty I series, have been plotted against depth in Fig. 10.

The chronological implications of the calibration seem to be of least importance and indeed any unexpected aspects of chronology postulated in this study have arisen from C^{14} dating rather than calibration. The suggested French connection of the earliest Neolithic settlers still remains plausible in spite of new dates approaching 5000 BC. The construction period of stone alignments and circles is now bracketed by the dates c 1600-c 850 BC which still means middle/late Bronze Age times in calibrated terms. The initiation of blanket peat in the study area is pushed back to c 1600 BC which still does not invalidate the local topographical factors and possible human influence which were considered as causal factors. By Iron Age times the narrowing gap between calibrated and uncalibrated dates means that no drastic chronological revisions will have to be made.

Once a reliable method of calibration is available, the implications for palynology will be of major importance. Already, the floating calibration established by Pearson et al., (1977) for the period 3600-4550 bp has shown that the bunching of radiocarbon dates around 4000 bp for the final disappearance of *Pinus* from Irish pollen diagrams, cannot be attributed to a calibration 'wiggle' and must, therefore represent a synchronous event. Whether or not the synchroneity of the elm decline throughout NW Europe, will also be established by calibration, remains to be seen.

Perhaps the most interesting implication of calibration is the possibility of tracing evolutionary patterns in the extended periods of early farming and metallurgy. The potential of such an approach has been outlined by Fleming (1975, 107) and in time, it may well produce interesting results in the study area which has a Neolithic span of about 2500 years (from c 5000-2500 BC) followed by a Bronze Age period of approximately 1700 years (c 2500-c. 750 BC). The earlier Neolithic clearance episodes are lengthened as a result of calibration and the discontinuity in late Neolithic times is further emphasized. The lengthening of the Bronze Age phases tends to reinforce the ideas already expressed concerning the gradual increase through time of invention/innovation rates as the society and economy increase in complexity. At the moment it seems that the real value of calibration will lie along these lines rather than in causing major chronological upheavals.

Table 7: Radiocarbon dates used in this study, with their calibrations according to Suess 1978. It should be noted that these calibrated dates are only intended to provide a general impression of the trend of calibration and therefore should not be used as absolute dates.

Laboratory No.	Radiocarbon Years bp	bc/ad	Calendar Years (after Suess 1978)	Sites
UB-2372	265±95	1685 ad	——	
UB-2371	1360±85	590 ad	493 AD	
UB-2370	2470±60	520 bc	766 BC	
UB-2414	3090±60	1140 bc	1322 BC	
UB-2368	3965±60	2015 bc	2508 BC	Cashelkeelty I
UB-2367	4915±95	2965 bc	3687 BC	
UB-2413	5845±100	3895 bc	4690 BC	
UB-2366	4035±85	2085 bc	2333 BC	
GrN-9171	800±30	1150 ad	1157 AD	
GrN-9172	2665±50	715 bc	863 BC	Cashelkeelty II
GrN-9173	2920±60	970 bc	1171 BC	
GrN-9279	1060±30	890 ad	913 AD	Cashelkeelty III
GrN-9174	2020±30	70 bc	125 BC	Dromatouk I
GrN-9346	3330±50	1380 bc	1609 BC	
GrN-9283	985±45	965 ad	1140 AD	Dromteewakeen I
GrN-9280	3265±55	1315 bc	1593 BC	Maughanasilly I
GrN-9281	3265±55	1315 bc	1593 BC	
GrN-9282	975±45	975 ad	1117 AD	Cullenagh I

CHAPTER 9

SUMMARY

This study is concerned with the identification and analysis of changes in the environmental systems of SW Ireland throughout prehistoric times. These changes are considered within the framework of regional environment and economic area, while at a more local level, the environmental setting of stone circle and alignment sites which belong to middle/late Bronze Age times, is examined. Archaeological evidence is then evaluated in conjunction with the palaeoecological data in order to build a model of prehistoric subsistence patterns.

Pollen analysis is the major tool used for demonstration and explanation of vegetational changes, supplemented by macrofossil and charcoal analysis. Archaeological excavation was carried out in order to gain more specific information regarding the sociocultural systems of the stone circle/alignment communities. Radiocarbon determinations are used widely in establishing the chronological framework.

The following is the proposed model of prehistoric subsistence patterns in SW Ireland:

<u>c 4000-3000 bc</u>

During this period both the hillslopes and valleys of SW Ireland were heavily wooded, with pine and oak forming the major woodland components. The predominant soils were forest brown earths and mild climatic conditions prevailed. The earliest evidence for human activity in the study area belongs to this millennium and it is postulated that migrating farming communities (possibly from NW France) were responsible for the limited woodland clearance which was carried out. Such population displacement is likely to have arisen as a result of pressure placed on natural resources by increasing population density in a well-established agricultural society. The subsistence pattern of these earlier Neolithic settlers is little understood but it is likely that the cereal cultivation and pastoralism were supplemented by exploitation of aquatic and forest resources. The settlement area of this period is unknown but a coastal emphasis is likely.

<u>c 3000-2000 bc</u>

Changes are evident in woodland composition during this time. Pine was gradually disappearing and was succeeded by oak, which together with birch and holly formed a dense climax woodland. The decline in elm values noted at c 3000 bc is equated with the classical elm decline of NW Europe and while human activity is ruled out as a possible causal factor, local climatic changes or more likely an epidemic fatal disease may have been responsible. General soil and climatic conditions in the study area remained unchanged from the preceding period.

The settlements established during the fourth millennium were deserted and no evidence was found to suggest human interference with the dense climax woodland. Changes in natural resources cannot be invoked to explain such a discontinuity in settlement and it is suggested that failure in some way to adapt to their environment made the early settlers extra susceptible to disasters such as disease, famine or crop failure.

c 2000-500 bc

The changes in vegetational cover from this time onwards were largely the result of human activity. All constituents of the mixed oak woodland were affected by clearances and pine finally disappeared at c 1000 bc. The area under grassland expanded rapidly, apparently bordered by hazel and ash scrub. Shortly after woodland clearance began, important changes took place in soil development. The process of podzolization began in many parts of the study area in the second millennium, by which time heathland conditions were spreading. In areas where local topographical conditions were favourable, the growth of blanket peat was initiated. Man's disturbance of the soil cover was a probable contributing factor to this initiation of peat growth. The climate apparently degenerated after c 1000 bc, resulting in cooler and wetter conditions.

The study area was a focus of settlement during these Bronze Age times. Movements of people from other parts of Ireland and both Britain and the Continent are envisaged and it is suggested that the main attractions offered by SW Ireland at this time were (i) new lands to be farmed and (ii) easily worked copper ore deposits. The subsistence pattern of these Bronze Age settlers must have rapidly acquired a stable food-producing character. Cereal cultivation and pastoralism became well-established and while artefacts such as barbed and tanged arrowheads are indicative of hunting, it has not been possible to evaluate its relative importance in the general economic system. It is probable that fishing also continued to be of importance.

Exploitation of the copper ores, which had begun by at least c 1500 bc, rapidly led to the establishment of a complex and extensive exchange network system. These industrial developments must in turn have triggered off changes in pre-existing economic and social systems. Population density increased, stable adjustment to the environment was achieved and formal burial/ritual monuments were constructed. Organization of the landscape first took place at this time, as evidenced by the appearance of field systems. On the basis of monument distribution, it is assumed that the area of settlement was concentrated along the coastal plains and around the headwaters of major rivers and their tributaries.

c 500 bc-300 ad

Woodland regeneration took place and hazel, birch and ash invaded abandoned clearings. Podzolization and gleying of the soils continued and blanket peat was spreading rapidly. A drastic reduction in human activity is evident and from what little evidence is available, emphasis on pastoralism is a possibility.

The reasons for this marked recession are sought in the widespread soil degeneration and spreading peat, which under the increasing population pressure of later Bronze Age times, must have ultimately caused a collapse of the agricultural system. This in turn, would have seriously undermined the industrial sector of the economy which also apparently declined. A real decline in population and fragmentation of society resulted, from which emerged the self-sufficient unit represented by the ringfort.

c 300 ad-present

Limited clearance of secondary woodland took place in Early Christian times and further changes are not evident until the thirteenth century, when it is suggested, the Normans introduced the exotic species beech and walnut. By the eighteenth century, planting of introduced species was well-established. Blanket peat continued to spread, as it still does in some places today.

Fluctuating episodes of farming activity may be traced throughout these centuries and in general, the subsistence pattern appears to have been one based on mixed farming, much as it is today. The area of settlement apparently shifted eastward onto the peat-free and more fertile brown earth lowlands, although during the eighteenth and nineteenth centuries severe population pressure meant that the marginal peatlands were again settled and farmed intensively.

This suggested scheme of prehistoric development can only be considered as an approximation. Evaluation of the reliability of these conclusions depends largely on the reliability of the empirical data and of the analytical methods employed. Severe limitations have been imposed on this study by the sheer lack of empirical archaeological and in particular archaeolozoological data. Certain interpretative problems inherent in palynological methodology have introduced additional uncertainties. I have no hestitation however, in presenting this model as the most probable at the present level of research. It attempts to explain the observed facts and allows suggestions to be made concerning the nature of the processes which influenced prehistoric societies in SW Ireland. Future work will have to concentrate on the accumulation and analysis of archaeological data and on more detailed environmental studies, before anything approaching a reliable model of prehistoric processes can be presented.

Postscript

The debate concerning the presence of late Neolithic activity in S Ireland (see above) has received some support from a recently determined C^{14} date of 4675 ± 35 bp or 2725 bc (GrN-9708) from Ardcrony, Co. Tipperary (Drs J Lanting, personal communication). The material dated was a long bone from the central cist of a cairn which on excavation, was found to contain two skeletons and a decorated pot (Wallace, 1977).

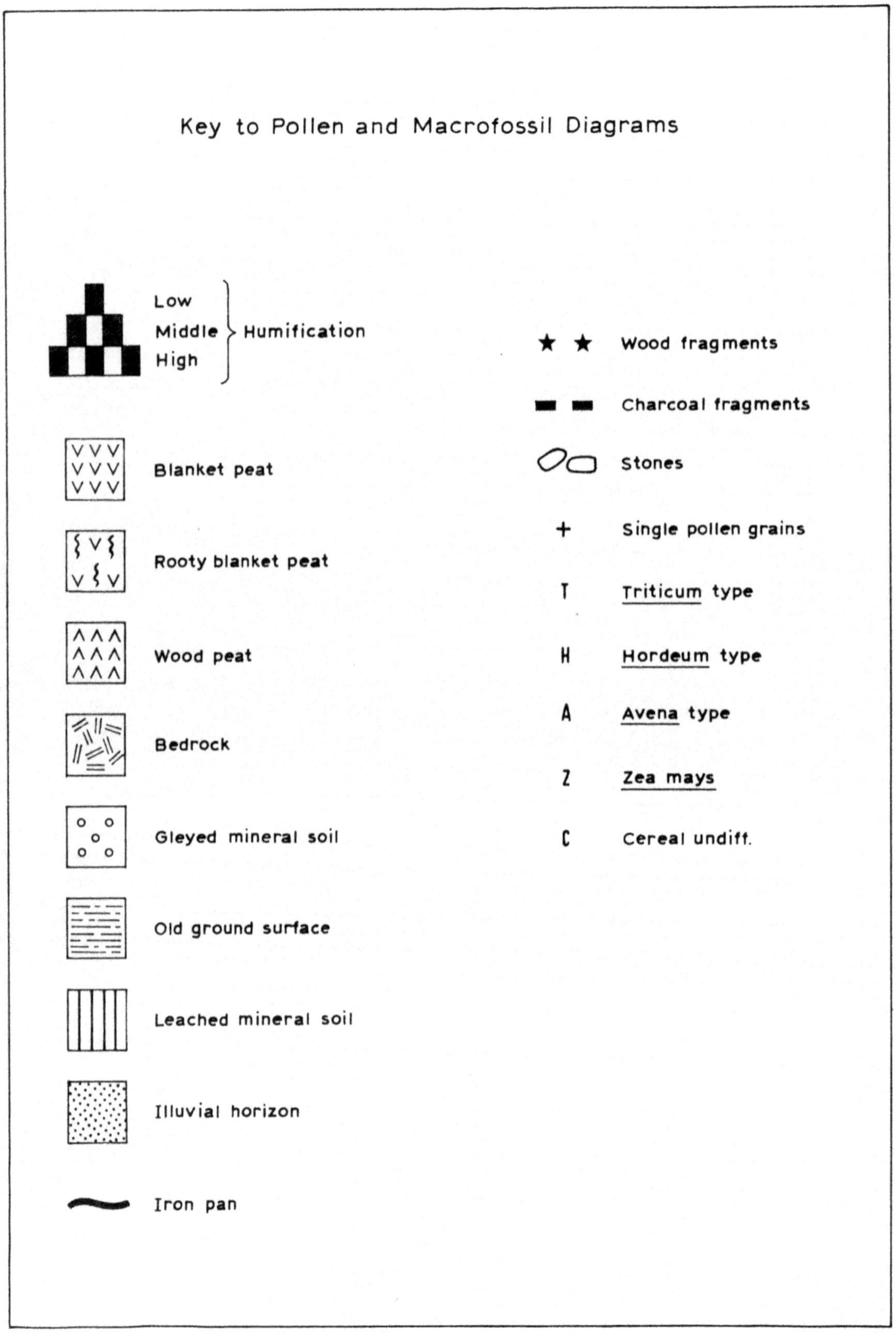
Key to Pollen and Macrofossil Diagrams
Low
Middle
High
Humification
Blanket peat
Rooty blanket peat
Wood peat
Bedrock
Gleyed mineral soil
Old ground surface
Leached mineral soil
Illuvial horizon
Iron pan
Wood fragments
Charcoal fragments
Stones
Single pollen grains
T Triticum type
H Hordeum type
A Avena type
Z Zea mays
C Cereal undiff.

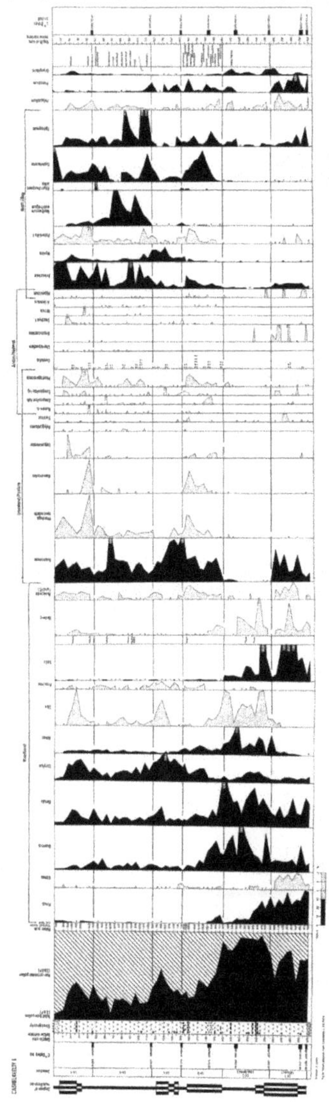

Fig 16

Please note that a full-size version of this figure is available to download from www.barpublishing.com/additional-downloads.html. The original foldout has been reduced in size to match the A4 format of this book, the image is therefore not as clear as the original foldout. Please refer to the original foldout via the download for the original content.

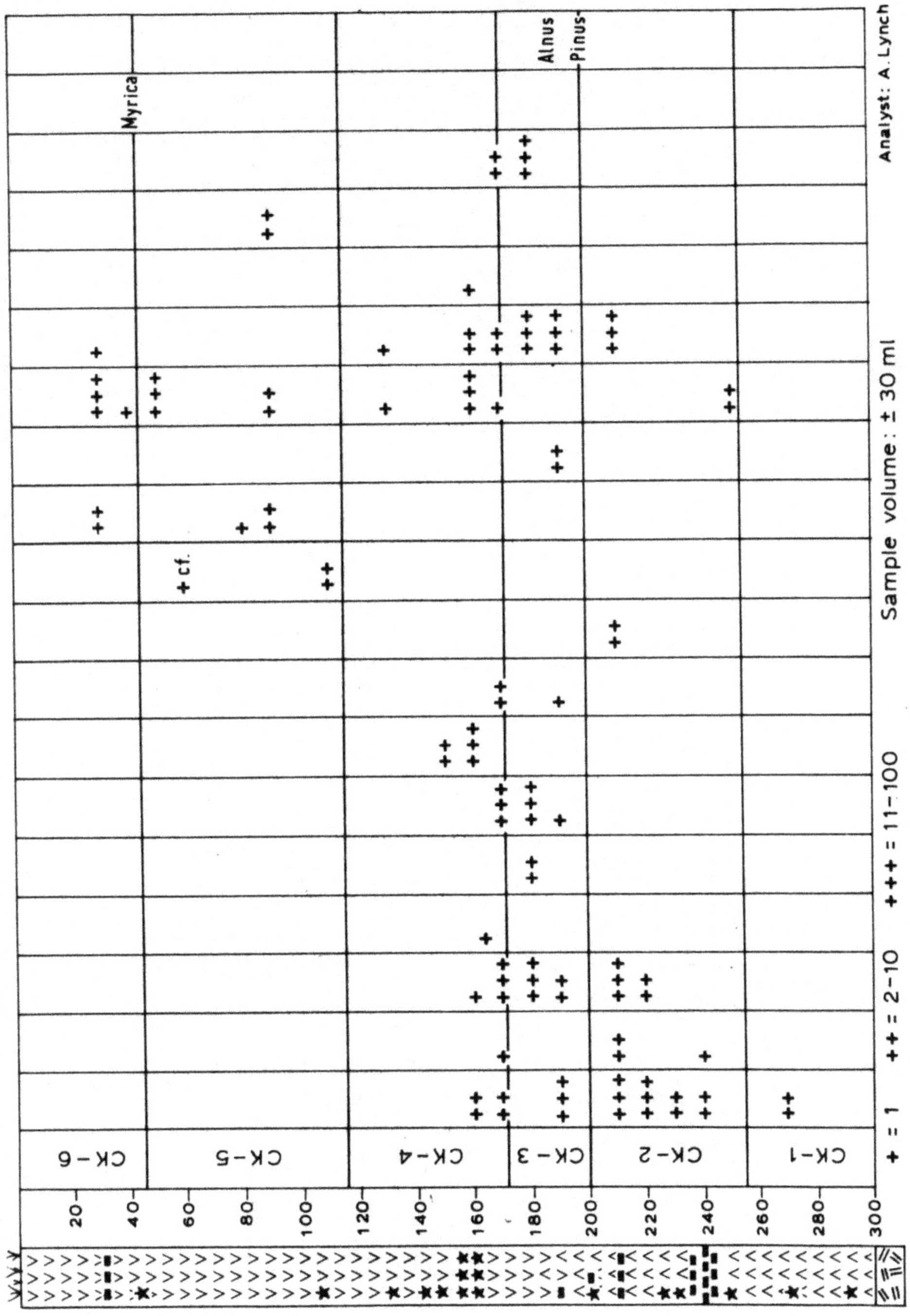

Fig.17 Cashelkeelty I macrofossil remains.

CASHELKEELTY II

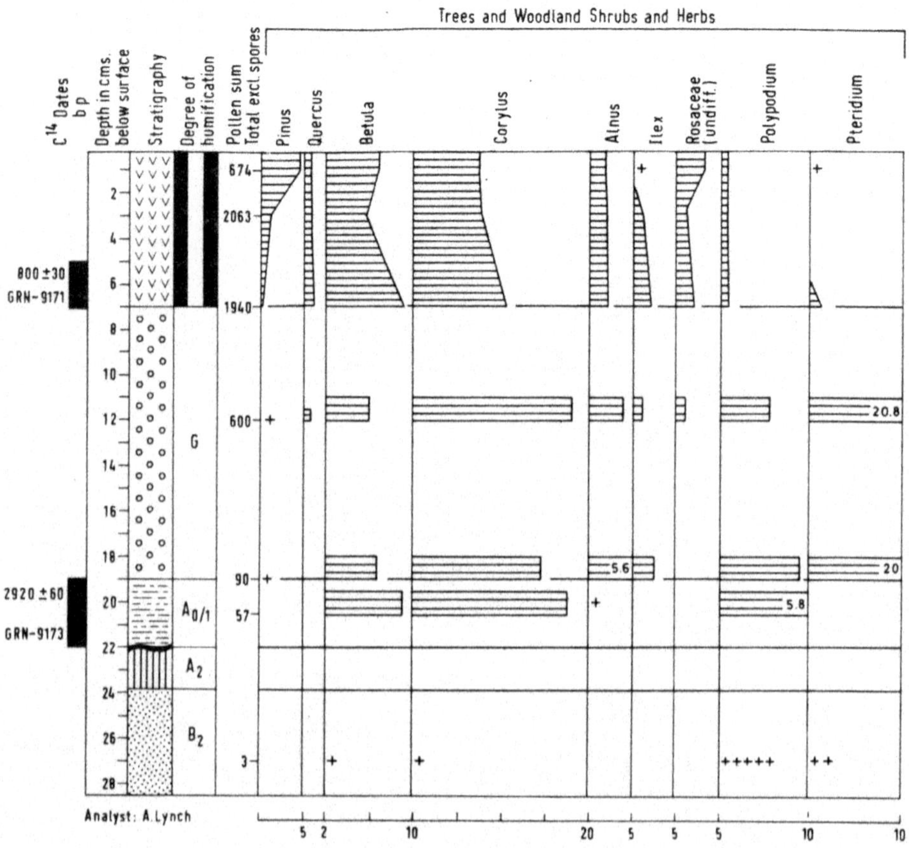

CASHELKEELTY III

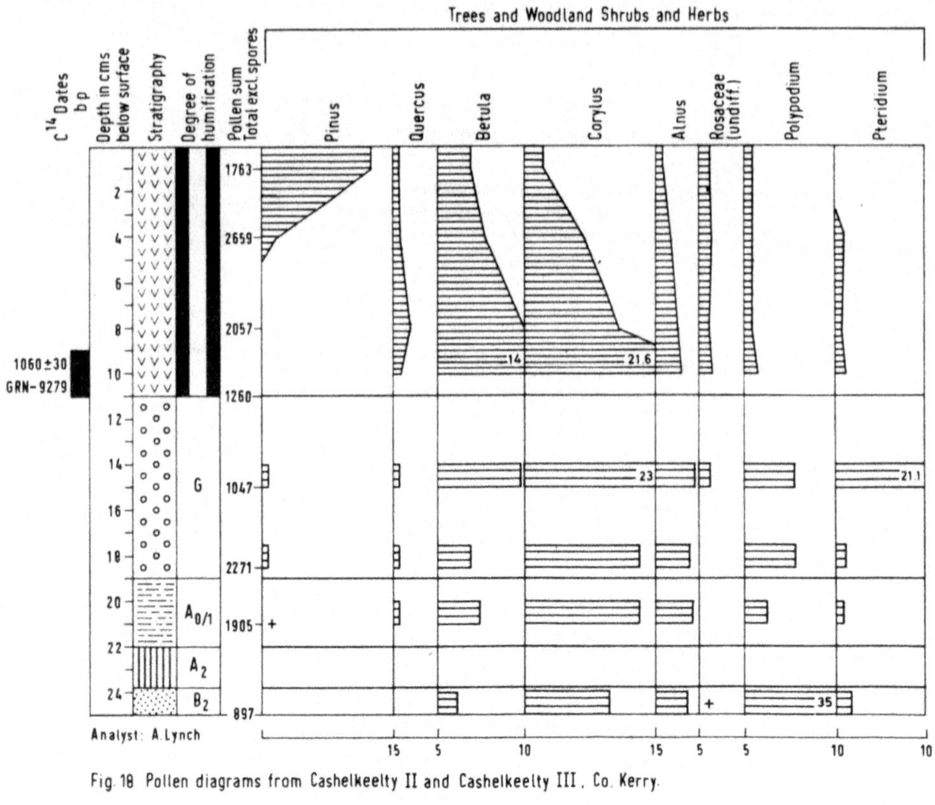

Fig. 18 Pollen diagrams from Cashelkeelty II and Cashelkeelty III, Co. Kerry.

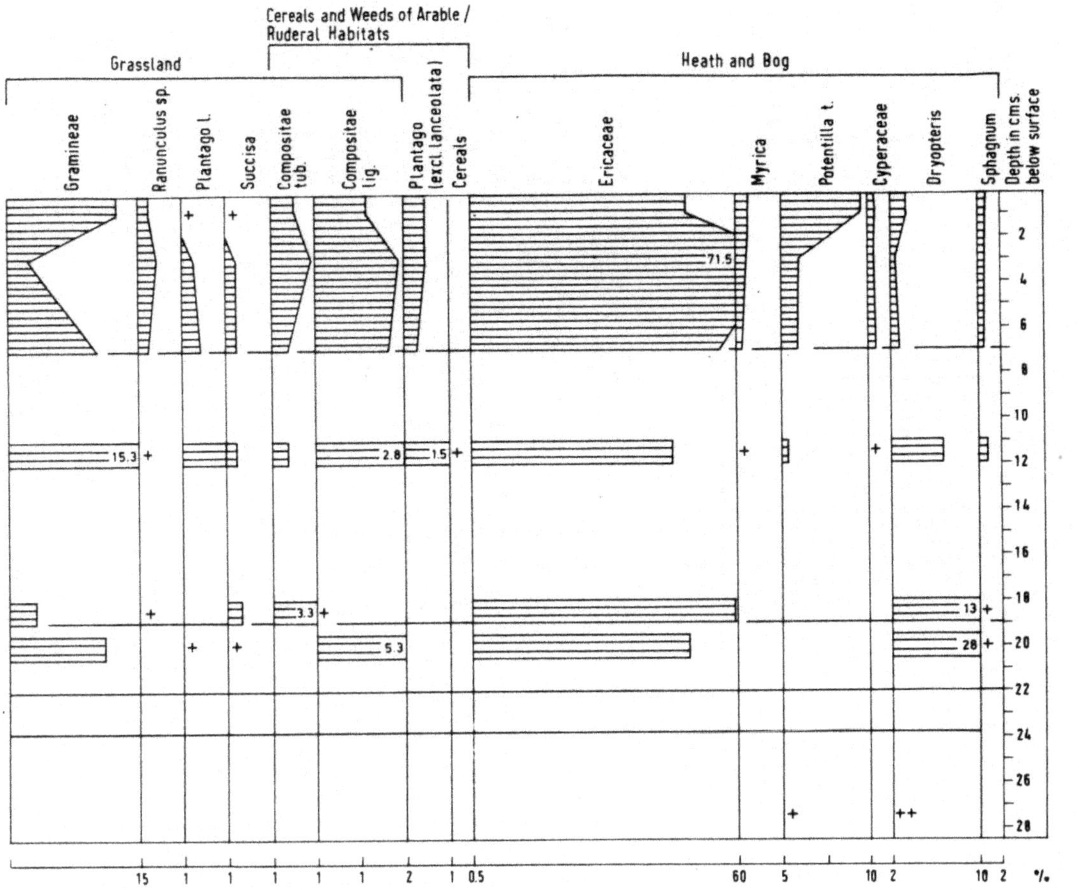
Cereals and Weeds of Arable / Ruderal Habitats
Grassland
Heath and Bog
Gramineae
Ranunculus sp.
Plantago l.
Succisa
Compositae tub.
Compositae lig.
Plantago (excl. lanceolata)
Cereals
Ericaceae
Myrica
Potentilla t.
Cyperaceae
Dryopteris
Sphagnum
Depth in cms. below surface
%

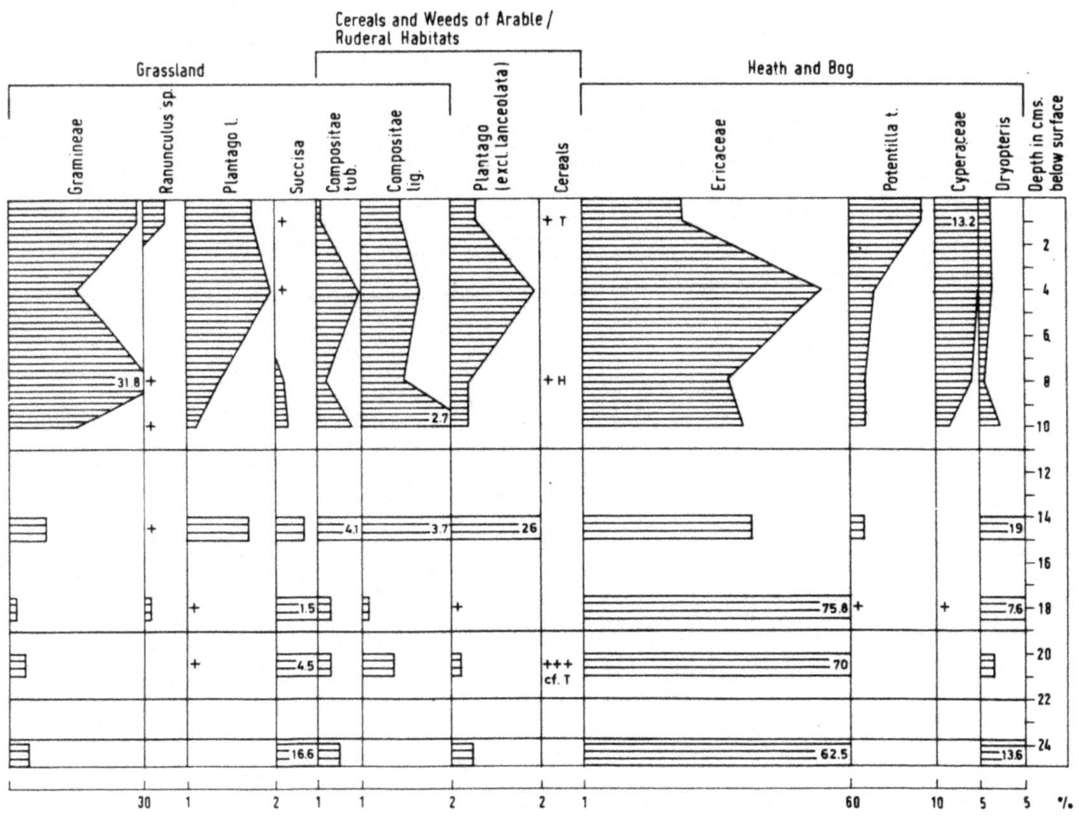
Cereals and Weeds of Arable / Ruderal Habitats
Grassland
Heath and Bog
Gramineae
Ranunculus sp.
Plantago l.
Succisa
Compositae tub.
Compositae lig.
Plantago (excl. lanceolata)
Cereals
Ericaceae
Potentilla t.
Cyperaceae
Dryopteris
Depth in cms. below surface
%

Table 8: Cashelkeelty II, Co. Kerry. Taxa not included in the pollen diagram, expressed as percentages of total pollen (ΣP).

Depth (cm)	Pollen Sum	Ulmus	Fraxinus	Hedera	Chenopodium	Rumex a. type	Leguminosae	Caryophyllaceae	Urtica	Polygalaceae	Galium	Narthecium	Hypericum	Eupotomageton
1	674	-	-	-	-	-	0.1	-	-	0.1	-	-	0.1	-
3	2063	-	-	0.1	0.1	0.1	0.1	-	0.1	-	0.1	0.1	0.1	-
7	1940	-	0.2	0.1	-	0.2	0.5	-	-	0.1	0.1	0.2	-	0.1
12	600	0.1	-	0.1	-	-	0.1	0.3	-	-	-	-	-	-
19	90	-	-	-	-	-	-	-	-	-	-	-	-	-
20.5	57	0.1	-	-	-	0.1	-	-	-	-	-	-	-	-
27	3	-	-	-	-	-	-	-	-	-	-	-	-	-

Table 9: Cashelkeelty III, Co. Kerry. Taxa not included in the pollen diagram, expressed as percentages of total pollen (ΣP)

Depth (cm)	Pollen Sum	Ulmus	Fraxinus	Salix	Ilex	Hedera	Leguminosae	Stachys t.	Urtica	Caryophyllaceae	Galium	Polygalaceae	Myrica	Sphagnum	Varia
1	1763	0.1	0.2	0.1	0.1	0.1	0.2	-	0.1	-	0.3	0.1	-	-	Fagus; Artemisia; Chenopodium; Rumex; Cruciferae; Pedicularis
4	2659	0.1	-	-	0.4	0.1	0.2	0.6	0.1	-	0.1	0.2	0.2	0.4	Narthecium; Hypericum; Eupotamogeton; Cruciferae
8	2057	0.1	-	0.2	0.6	0.3	0.1	0.1	-	-	0.1	-	0.6	0.4	Narthecium; Hypericum; Eupotamogeton
11	1260	-	-	0.1	1.5	0.2	0.1	0.3	-	-	-	-	0.6	0.4	
15	1047	0.4	-	0.1	0.2	-	0.4	-	-	-	-	-	0.2	0.1	
18	2271	-	-	-	0.1	-	0.1	-	-	0.1	-	-	-	-	
21	1905	-	-	-	0.1	-	-	-	-	0.1	-	-	-	0.1	
25	595	-	-	-	0.2	-	-	-	-	-	-	-	-	0.8	Artemisia

CASHELKEELTY IV

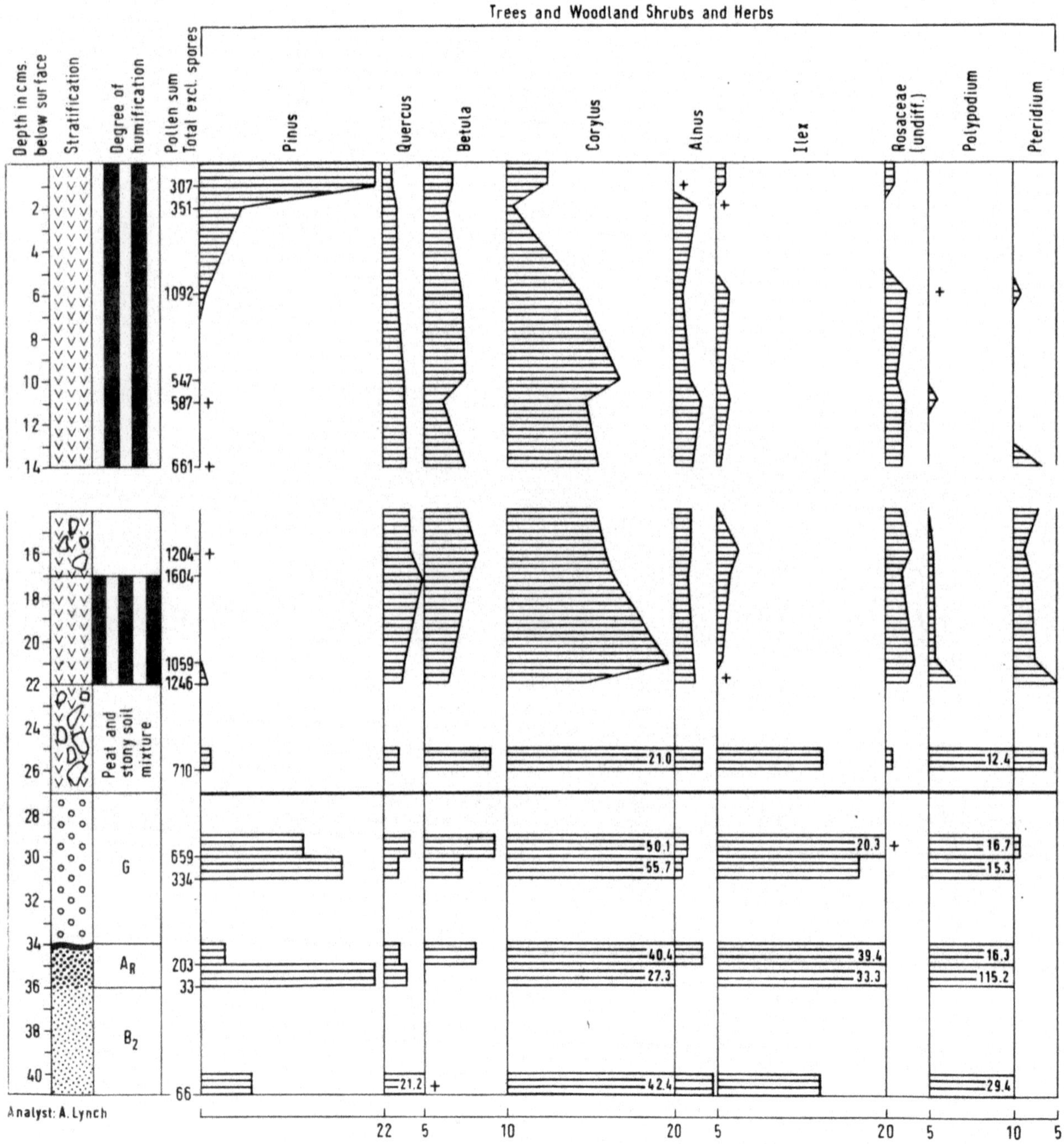

Fig. 19 Pollen diagram from Cashelkeelty IV, Co. Kerry.

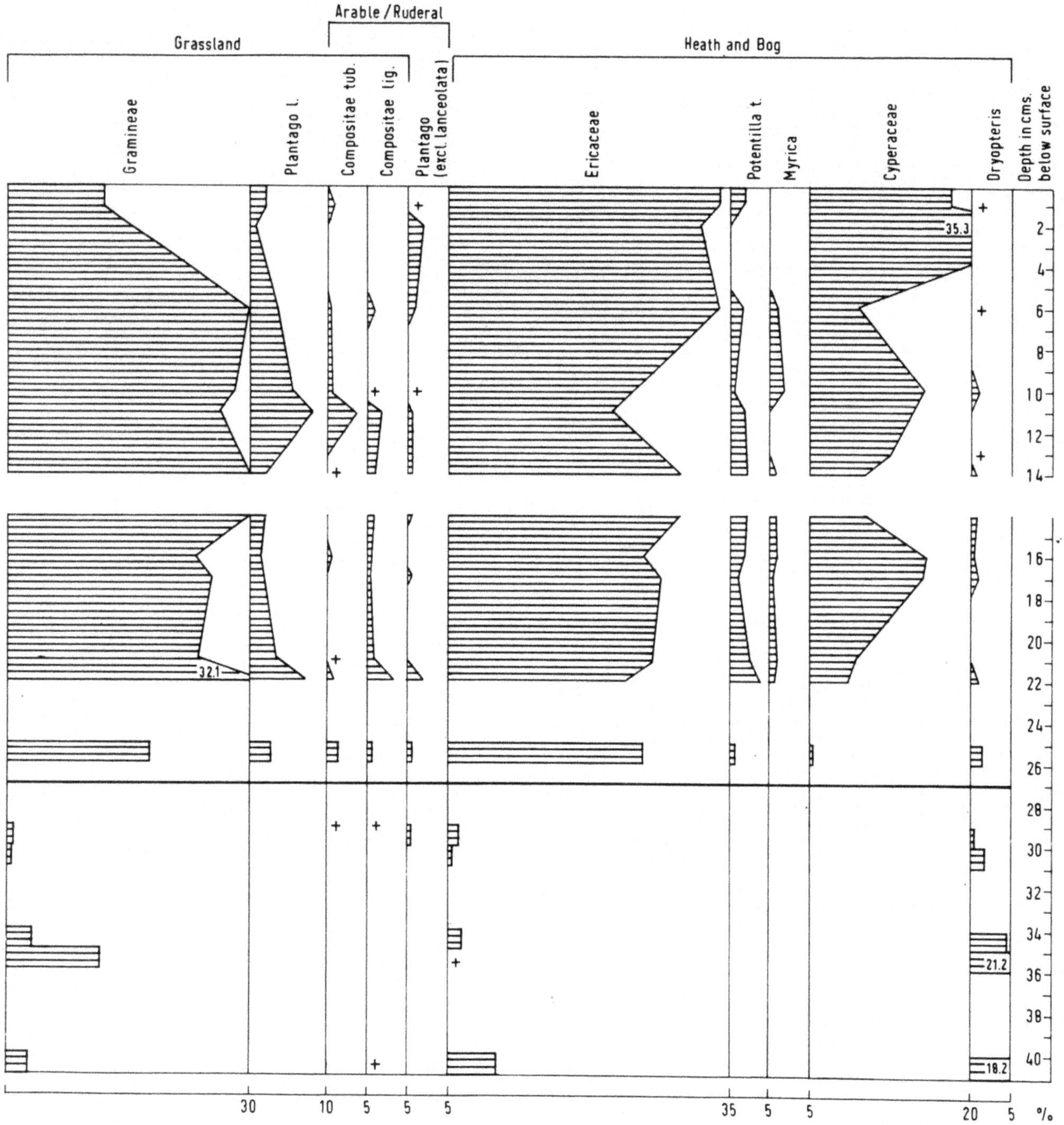

Arable / Ruderal
Grassland
Heath and Bog
Gramineae
Plantago l.
Compositae tub.
Compositae lig.
Plantago (excl. lanceolata)
Ericaceae
Potentilla t.
Myrica
Cyperaceae
Dryopteris
Depth in cms. below surface
35.3
32.1
21.2
18.2
30
10
5
5
5
35
5
5
20
5
%

Table 10: Cashelkeelty IV, Co. Kerry. Taxa not included in pollen diagram, expressed as percentages of total pollen (ΣP).

Depth (cm)	Pollen Sum	Salix	Fraxinus	Hedera	Ranunculus	Succisa	Leguminosae	Stachys t.	Campanulaceae	Urtica	Rumex a. type	Artemisia	Chenopodium	Sphagnum	Varia
1	307	-	-	-	-	-	-	-	-	-	-	-	0.3	-	
2	351	0.3	0.9	-	-	-	-	-	-	-	0.6	0.3	-	0.6	Ulmus; Fagus
6	1092	0.2	0.1	-	0.3	0.1	0.5	0.8	-	-	-	0.1	-	0.5	
10	547	0.2	-	-	-	-	0.4	-	-	0.5	0.4	-	-	0.4	Ulmus; Galium
11	587	-	-	-	0.7	0.2	2.0	3.1	-	-	-	-	0.2	0.2	Spergula
14	661	0.5	0.6	0.2	0.3	-	0.3	1.8	-	0.5	0.2	-	-	0.2	
16	1204	0.4	-	0.1	0.7	-	-	3.3	0.1	-	0.1	0.1	-	0.3	Ulmus
17	1604	0.1	0.2	-	-	0.1	-	1.9	-	0.1	-	-	-	-	
21	1059	0.1	-	-	0.1	0.1	-	5.7	0.1	-	-	-	-	0.2	
22	1246	-	0.1	0.1	-	0.5	0.1	4.2	0.1	-	-	-	-	0.3	Spergula a.
26	710	-	-	-	0.3	0.1	0.7	0.1	-	-	-	-	-	0.8	Umbelliferae; Spergula a.
30	659	-	-	0.3	-	0.2	-	0.9	-	-	-	-	-	-	
31	334	-	0.3	0.6	-	-	-	-	-	-	-	-	-	-	
35	203	-	-	-	-	-	-	-	-	-	-	-	-	-	
36	33	-	-	-	-	-	-	-	-	-	-	-	-	-	
41	66	1.5	-	-	-	-	-	-	-	-	-	-	-	-	

DROMATOUK I

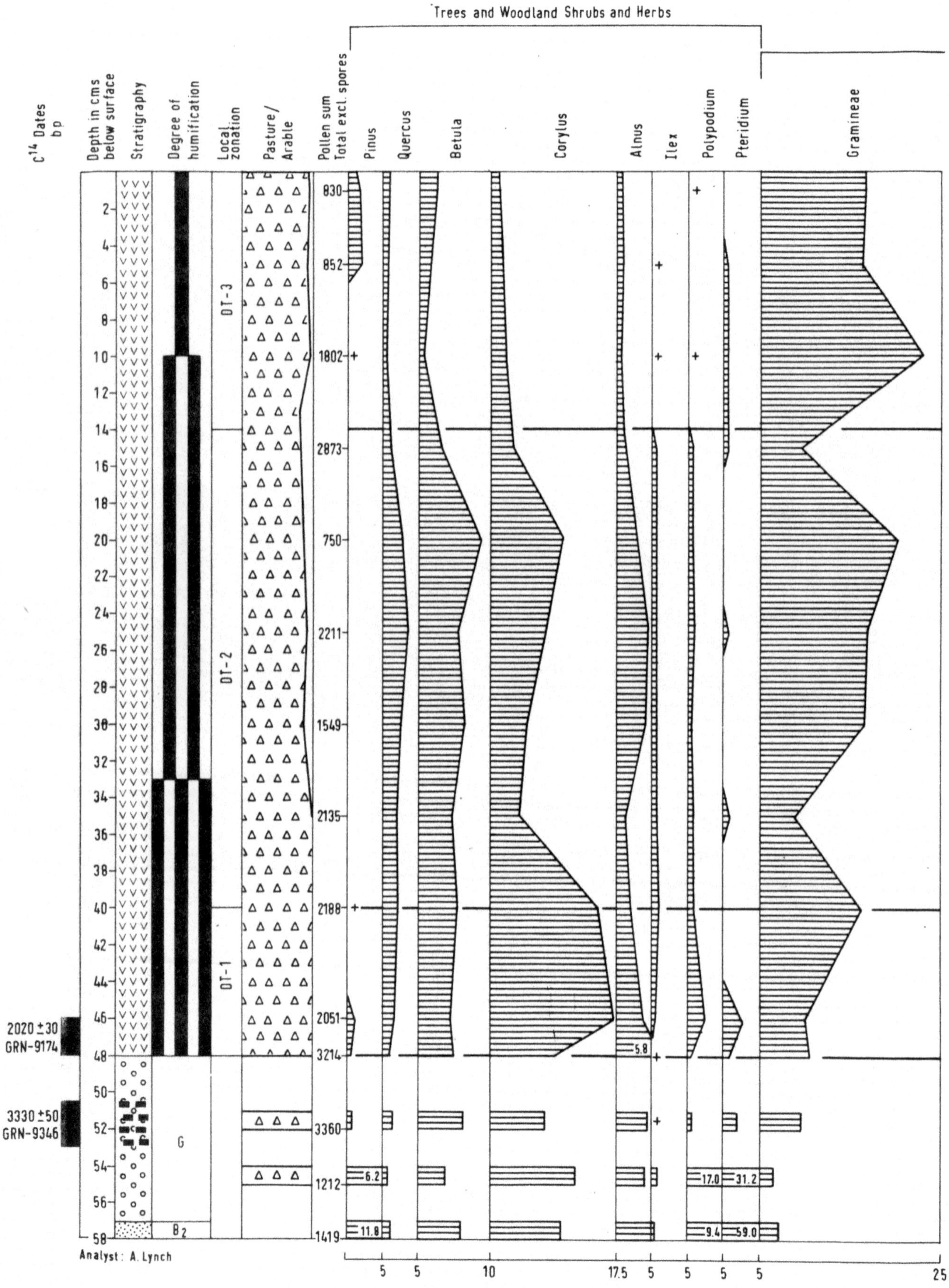

Fig. 20 Pollen diagram from Dromatouk I, Co. Kerry.

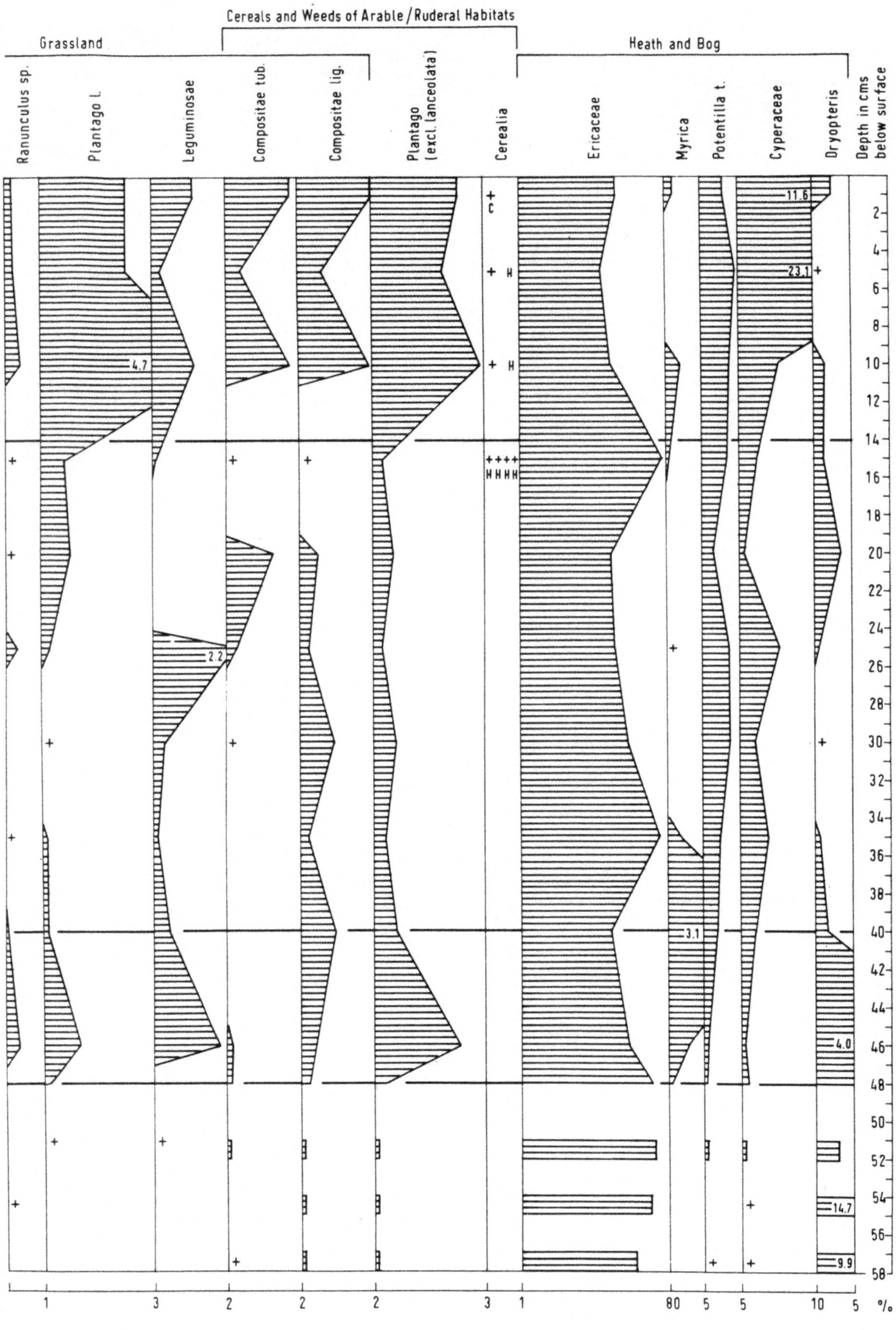

Cereals and Weeds of Arable/Ruderal Habitats
Grassland
Heath and Bog
Ranunculus sp.
Plantago l.
Leguminosae
Compositae tub.
Compositae lig.
Plantago (excl. lanceolata)
Cerealia
Ericaceae
Myrica
Potentilla t.
Cyperaceae
Dryopteris
Depth in cms below surface
%

Table 11: Dromatouk I, Co. Kerry. Taxa not included in the pollen diagram, expressed as percentages of total pollen (ΣP)

Depth (cm)	Pollen Sum	Ulmus	Fraxinus	Salix	Hedera	Rumex a. type	Artemisia	Rosaceae (undiff.)	Stachys t.	Umbelliferae	Caryophyllaceae	Hypericum	Narthecium	Galium	Sphagnum	Succisa	Pedicularis	Campanulaceae	Polygalaceae	Varia
1	830	0.1	0.1	-	0.1	0.4	1.0	-	-	-	-	0.1	-	0.1	-	-	-	-	-	
5	852	0.1	0.1	0.1	0.1	0.9	-	-	0.5	0.1	0.1	0.1	-	0.4	-	-	-	-	-	Cruciferae
10	1802	0.1	-	-	0.1	0.3	0.1	0.4	0.7	-	0.2	0.4	0.3	0.2	0.3	0.2	0.1	0.1	0.1	
15	2873	-	0.1	-	0.1	-	0.1	0.1	0.3	0.1	-	0.1	0.1	0.1	0.0	-	-	-	-	
20	750	0.1	0.4	-	0.3	-	-	0.3	0.5	-	0.1	-	-	-	-	-	-	-	0.3	
25	2211	0.3	0.7	-	0.1	0.1	0.1	0.1	0.7	0.4	0.3	0.1	-	-	0.1	-	0.2	-	-	
30	1549	-	0.3	-	-	0.1	-	-	2.5	0.2	0.1	-	-	-	0.1	-	0.3	0.1	-	
35	2135	-	0.1	-	-	0.1	-	-	1.3	-	-	-	0.1	-	0.1	0.1	0.1	-	0.1	
40	2188	-	0.1	-	0.1	-	-	-	2.8	-	-	-	-	-	0.1	-	0.4	0.1	-	
46	2051	0.1	0.1	-	0.1	-	-	-	0.5	-	-	-	-	-	0.2	-	-	-	-	
48	3214	0.2	0.1	-	0.1	-	-	-	0.1	-	-	-	-	-	-	0.1	0.1	-	-	
52	3360	0.3	0.1	-	0.1	-	-	-	0.2	-	-	-	-	-	0.1	0.1	-	0.1	-	
55	1212	-	-	-	0.2	-	-	-	-	-	-	-	-	-	0.8	0.1	-	-	-	
58	1419	-	-	-	0.2	-	-	-	-	-	-	-	-	-	0.4	-	-	-	-	Tilia

DROMTEEWAKEEN I

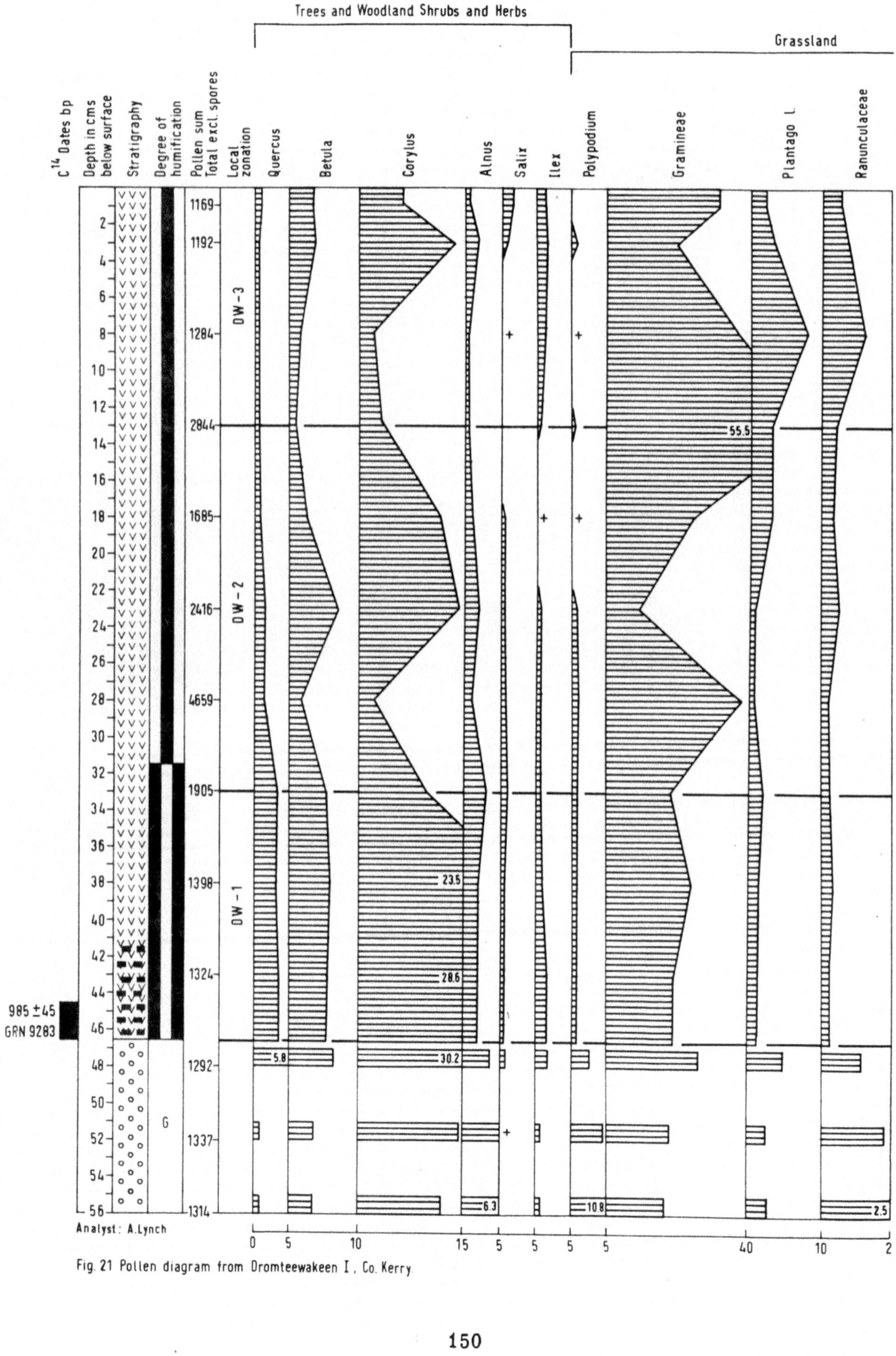

Fig. 21 Pollen diagram from Dromteewakeen I, Co. Kerry.

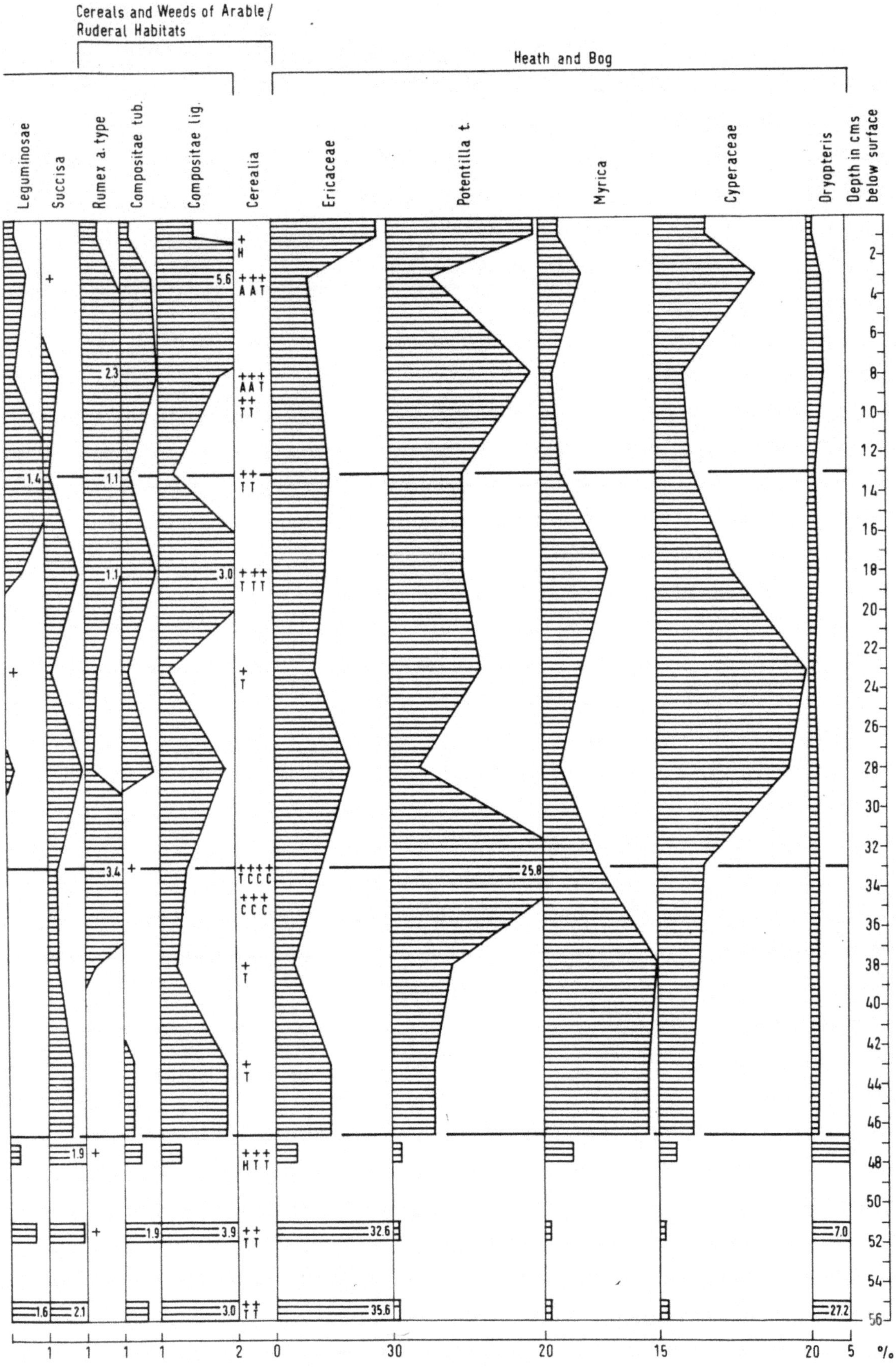
Cereals and Weeds of Arable/ Ruderal Habitats
Heath and Bog
Leguminosae
Succisa
Rumex a. type
Compositae tub.
Compositae lig.
Cerealia
Ericaceae
Potentilla t.
Myrica
Cyperaceae
Dryopteris
Depth in cms below surface
%

Table 12: Dromteewakeen I, Co. Kerry. Taxa not included in pollen diagram, expressed as percentages of total pollen (ΣP)

Depth (cm)	Pollen Sum	Pinus	Ulmus	Fraxinus	Artemisia	Chenopodium	Plantaginaceae	Rosaceae	Cruciferae	Stachys t.	Umbelliferae	Caryopyllaceae	Hedera	Hypericum	Narthecium	Galium	Pedicularis	Campanulaceae	Polygalaceae	Fagus	Pteridium	Sphagnum
1	1169	1.6	-	0.1	-	-	0.4	1.6	0.2	-	-	-	-	-	-	-	-	-	-	0.1	-	-
3	1192	0.3	-	0.4	0.4	0.2	2.9	-	-	-	0.4	-	0.3	0.3	-	0.3	0.3	-	0.1	-	0.1	0.1
8	1284	0.9	-	0.1	-	0.1	4.6	-	0.1	0.7	0.1	0.1	0.1	0.3	-	3.8	-	-	-	0.1	0.6	-
13	2844	-	0.1	-	-	-	2.0	-	-	1.4	-	-	-	0.1	-	0.4	-	0.1	-	0.1	0.2	0.1
18	1685	0.1	-	0.1	0.1	-	2.6	0.1	-	1.7	0.1	-	-	0.1	-	1.4	-	-	0.1	0.1	0.8	0.2
23	2416	-	-	0.2	-	-	1.0	0.1	-	-	0.1	-	0.1	-	0.1	-	-	-	-	-	-	0.1
28	4659	-	-	-	-	-	0.5	0.1	-	-	0.9	-	0.1	-	-	0.1	0.2	-	-	-	-	-
33	1905	-	-	0.4	-	-	1.3	0.2	-	0.1	0.1	0.1	0.3	-	0.3	0.1	-	0.1	-	0.8	0.1	0.1
38	1398	-	-	0.4	-	-	1.9	0.1	-	0.1	-	-	0.2	0.1	0.2	-	-	0.1	-	-	0.4	0.1
43	1324	-	0.1	0.2	0.1	-	2.3	-	-	0.2	-	-	0.2	-	-	0.1	-	-	-	-	-	0.1
48	1292	0.1	0.1	0.5	-	-	4.8	-	-	0.1	0.2	0.2	1.2	0.2	0.1	0.3	-	-	-	-	0.3	3.9
52	1337	-	-	-	0.2	-	2.1	-	-	-	-	-	-	-	-	-	-	-	-	-	0.7	1.7
56	1314	0.1	-	-	-	-	8.4	-	0.3	0.1	-	1.2	0.6	-	0.1	-	-	-	-	-	1.4	1.4

VARIA: Carpinus (1 cm); Epilobium (3 cm); Acer (18 cm); Urtica (28 cm); Juglans (33 cm); Mercurialis (48 cm); Polygonum (52 cm & 56 cm)

MAUGHANASILLY I

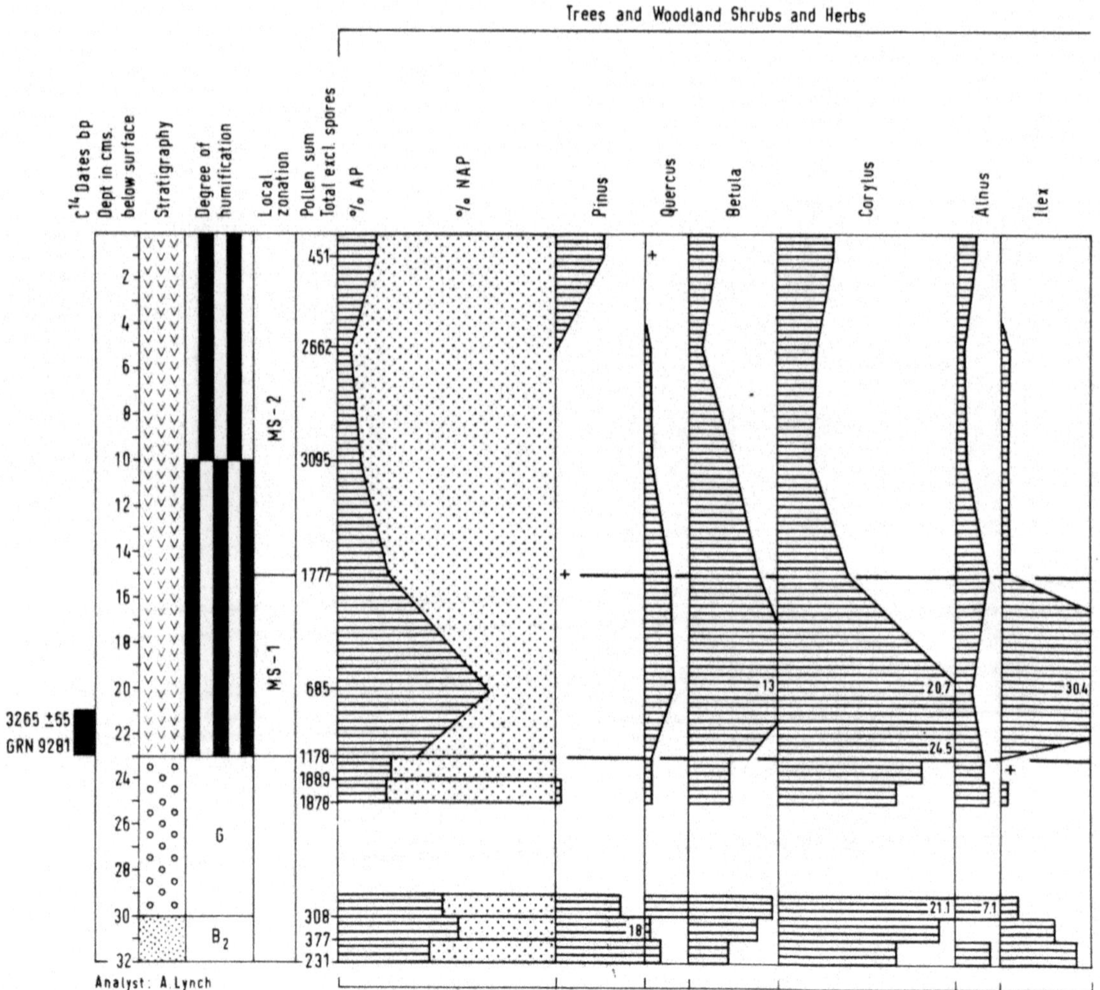

Fig. 22 Pollen diagram from Maughanasilly I, Co. Cork.

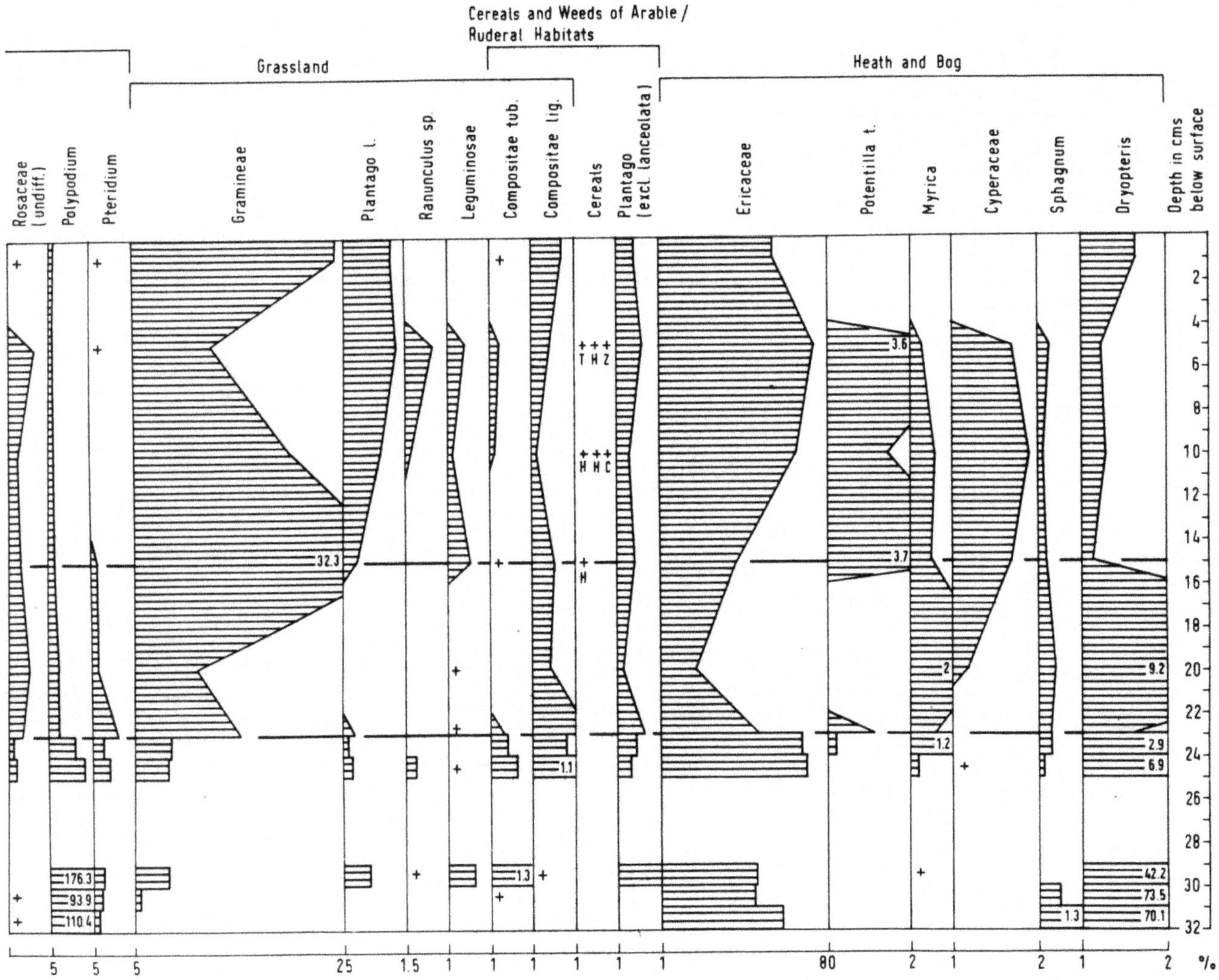
Cereals and Weeds of Arable / Ruderal Habitats
Grassland
Heath and Bog
Rosaceae (undiff.)
Polypodium
Pteridium
Gramineae
Plantago l.
Ranunculus sp.
Leguminosae
Compositae tub.
Compositae lig.
Cereals
Plantago (excl. lanceolata)
Ericaceae
Potentilla t.
Myrica
Cyperaceae
Sphagnum
Dryopteris
Depth in cms below surface
%

Table 13: Maughanasilly I, Co. Cork. Taxa not included in the pollen diagram, expressed as percentages of total pollen ($\sqrt{\Sigma}$P)

Depth (cm)	Pollen Sum	Ulmus	Hedera	Stachys t.	Salix	Chenopodium	Succisa	Urtica	Hypericum	Polygalaceae	Narthicium	Galium	Potamogeton	Varia
1	451	-	2.0	-	-	-	-	-	-	-	-	-	-	Fraxinus; Cruciferae
5	2662	0.1	0.1	0.1	-	-	-	0.1	0.2	0.1	0.1	0.2	-	
10	3095	-	0.2	0.2	0.1	0.1	0.1	0.1	-	-	0.1	0.1	0.1	
15	1777	0.1	0.1	0.6	0.2	-	-	-	-	0.1	0.1	-	0.3	
20	685	0.1	0.9	1.2	-	-	-	-	-	-	-	-	-	
23	1178	-	0.3	0.2	-	-	1.2	-	0.2	-	-	0.1	-	
24	1889	-	0.1	-	-	-	0.6	-	-	-	-	-	-	Artemisia
25	1878	0.1	0.1	-	-	0.1	0.4	-	0.1	-	-	-	-	
30	308	-	-	-	-	-	1.0	-	-	-	-	-	-	
31	377	-	-	-	-	-	0.5	-	-	-	-	-	-	
32	231	-	-	-	-	-	-	-	-	-	-	-	-	Pedicularis

CULLENAGH I

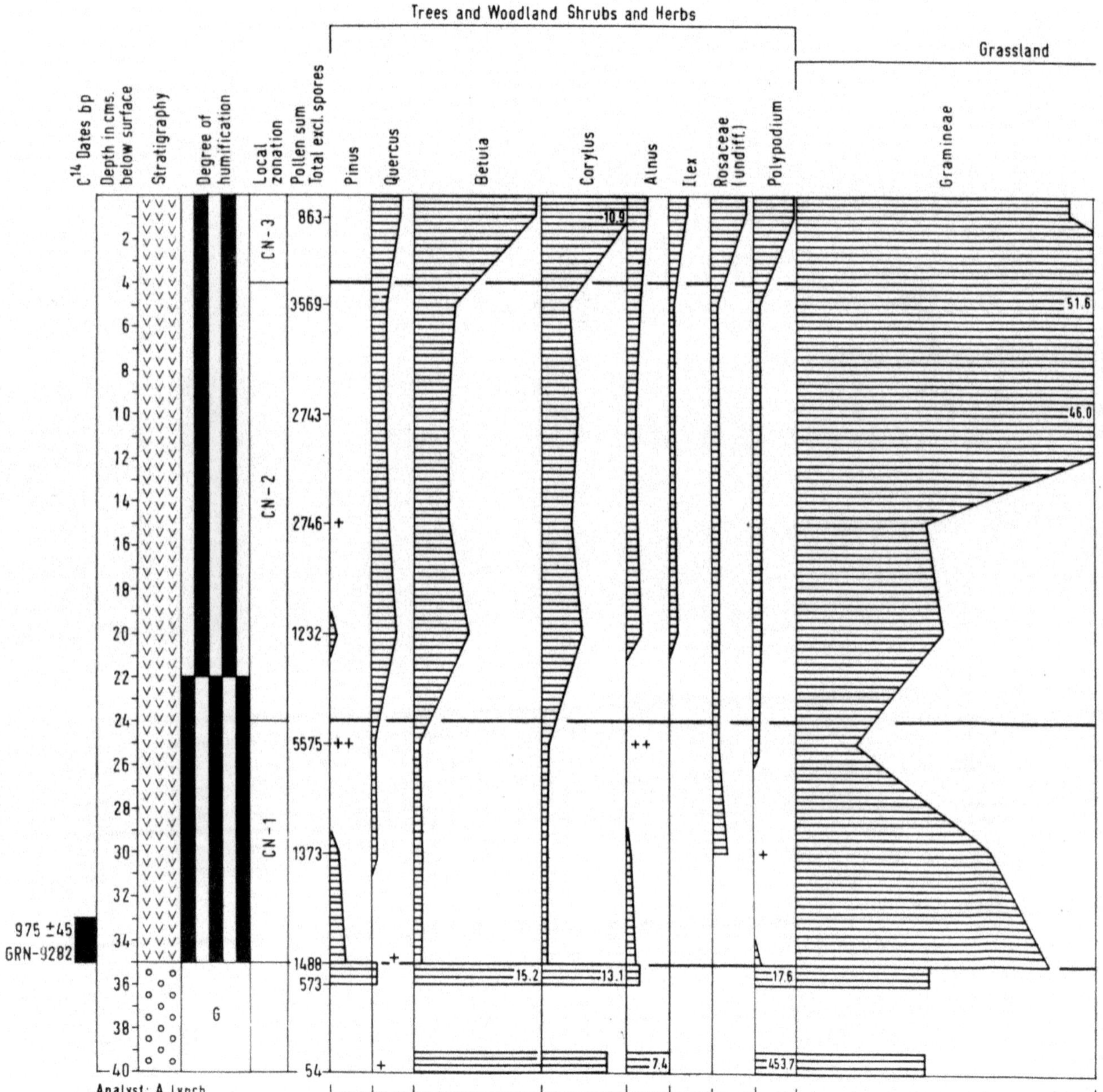

Fig. 23 Pollen diagram from Cullenagh I, Co. Cork.

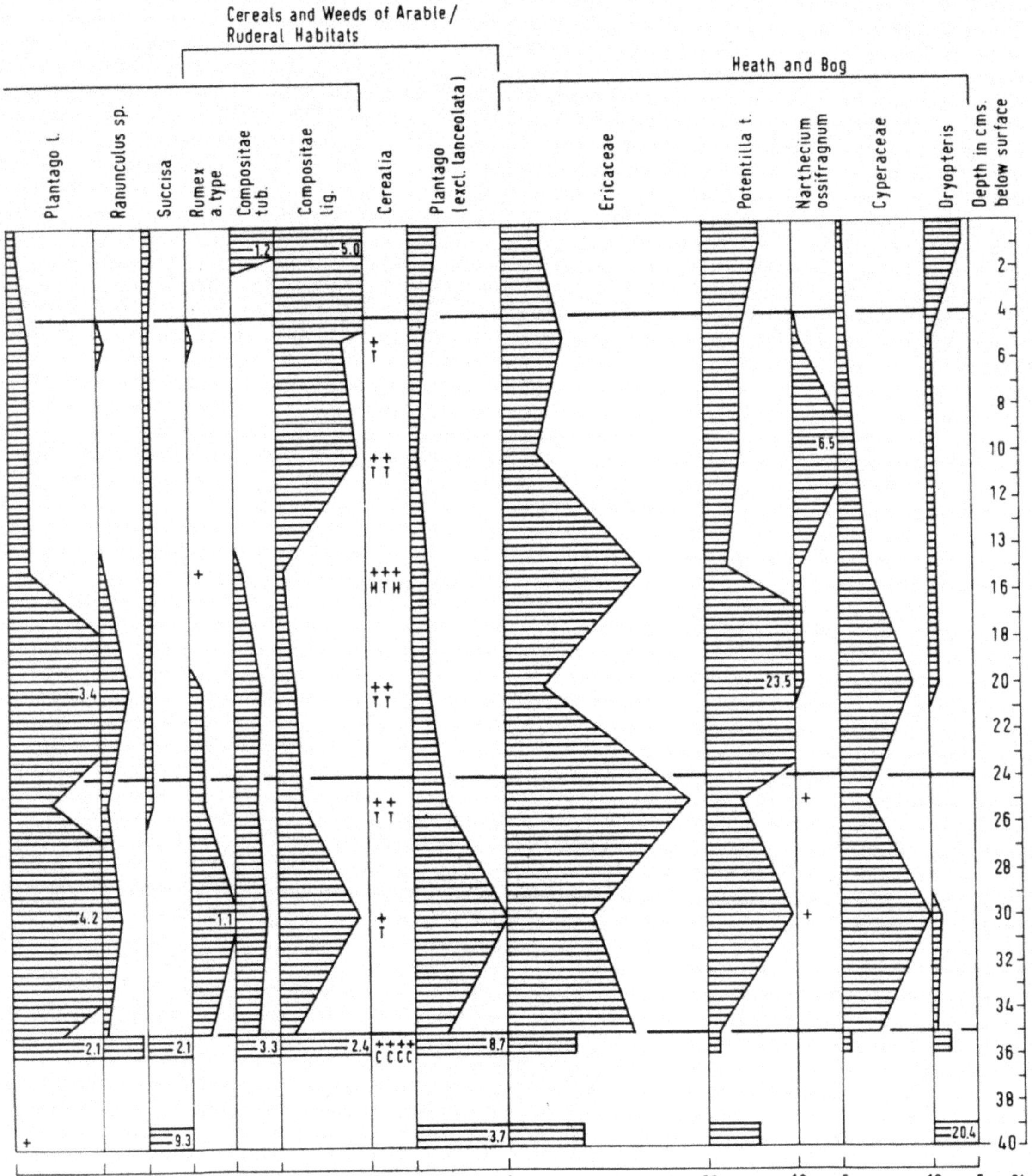

Cereals and Weeds of Arable/
Ruderal Habitats
Heath and Bog
Plantago l.
Ranunculus sp.
Succisa
Rumex a. type
Compositae tub.
Compositae lig.
Cerealia
Plantago (excl. lanceolata)
Ericaceae
Potentilla t.
Narthecium ossifragnum
Cyperaceae
Dryopteris
Depth in cms. below surface
%

Table 14: Cullenagh I, Co. Cork. Taxa not included in the pollen diagram, expressed as percentages of total pollen (Σ P)

Depth (cm)	Pollen Sum	Ulmus	Fraxinus	Salix	Juglans	Pteridium	Artemisia	Hedera	Cruciferae	Stachys t.	Leguminosae	Hypericum	Polygalaceae	Myrica	Pedicularis	Galium	Sphagnum	Campanulaceae	Umbelliferae	
1	863	-	-	0.1	-	0.2	-	-	-	1.0	-	-	-	-	-	-	-	-	-	
5	3569	0.1	0.3	-	-	0.1	0.1	0.1	-	2.2	-	-	0.1	0.2	0.2	-	-	0.1	-	Carpinus
10	2743	-	0.1	0.1	-	-	-	-	-	11.6	-	0.1	0.1	0.1	0.4	-	-	-	-	
15	2746	-	-	0.1	0.1	0.2	0.1	-	0.1	6.0	0.1	-	-	0.2	0.1	-	0.1	-	0.1	
20	1232	-	-	0.2	-	0.8	0.2	0.2	-	9.1	-	0.1	0.4	-	0.2	0.5	-	0.1	0.2	Urtica; Euonymus Europaeus
25	5575	-	-	0.1	-	0.1	0.1	-	-	0.6	-	-	0.1	-	-	0.4	0.1	-	0.1	
30	1373	0.1	-	-	0.4	0.3	-	-	-	2.5	0.7	0.3	0.1	-	-	0.4	-	-	-	
35	1488	-	0.1	0.1	-	-	-	-	0.3	0.4	0.3	0.1	-	-	-	0.3	-	-	-	
36	573	-	-	0.2	-	0.2	-	-	0.2	0.5	-	-	-	-	-	-	0.3	-	-	
40	54	-	-	-	-	-	-	-	-	-	-	-	-	-	-	-	-	-	-	

Appendix

Glossary of Latin Plant Names and their English Equivalents
(after Webb, 1977)

Adiantum capillus-veneris, maidenhair fern
Alnus, alder
Arbutus unedo, strawberry tree
Artemisia, mugwort
Avena, oats
Betula pubescens, brown birch
Callitriche intermedia, starwort
Calluna vulgaris, ling
Carex acuta, slender-spiked sedge
Chenopodium, goosefoot
Corylus, hazel
Crataegus monogyna, hawthorn
Cyperaceae, sedges
Daboecia cantabrica, St. Dabeoc's heath
Dryopteris, fern
Elatine hexandra, waterwort
Eleocharis multicaulis, many-stemmed spike-rush
Empetrum, crowberry
Ericaceae, heathers
Erica ciliaris, ciliate heath
Erica erigena, mediterranean heath
Erica mackaiana, mackay's heath
Erica tetralix, cross-leaved heath
Eriocaulon aquaticum, pipe-wort
Eriophorum angustifolium, cotton grass
Euphorbia hyberna, Irish spurge
Fagus, beech
Fraxinus, ash
Galium boreale, northern bedstraw
Gramineae, grasses
Hedera helix, ivy
Hordeum var nudum, naked barley
Hypericum canadense, St. John's wort
Ilex aquifolium, holly
Isoetes lacustris, quillwort
Juglans, walnut
Juncus articulatus, jointed rush
Juncus bulbosus, bulbous rush
Juncus effusus, common rush
Juniperus sibirica, juniper
Lavatera, mallow
Linum usitatissimum, flax
Lobelia dortmanna, water lobelia
Luzula sylvatica, great wood-rush
Molinia caerulea, purple moor-grass
Myrica gale, bog-myrtle
Najas flexilis, flexible naiad
Narthecium ossifragum, bog asphodel
Neotinea intacta, Irish orchid
Otanthus maritimus
Pinguicula grandiflora, greater butterwort
Pinguicula lusitanica, pink butter-wort
Picea sitchensis, sitka spruce
Pinus sylvestris, scot's pine
Plantago coronopus, buck's-horn plantain
Plantago lanceolata, ribwort plantain
Plantago maritima, sea plantain
Polygalaceae, milkwort
Polygonum lapathifolium, pale persicaria
Polypodium, fern
Potamogeton gramineus, various-leaved pondweed
Potentilla erecta, tormentil
Prunus spinosa, blackthorn
Pseudotsuga taxifolia, douglas fir
Pteridium aquilinum bracken
Quercus sessiliflora, sessile oak

Ranunculus flammula, lesser spearwort
Ranunculus lingua, great spearwort
Rhynchospora alba, white beak-sedge
Rumex acetosa t., sorrel
Salix cinerea, fen sallow
Saxifraga hirsuta, kidney saxifrage
Saxifraga spathularis, St. Patrick' s cabbage
Scilla nonscripta, squill
Schoenus nigricans, black bog rush
Scirpus caespitosus, deer grass
Scirpus tabernaemontani, glaucus club-rush
Sedum roseum, roseroot
Sibthorpia europaea, cornish moneywort
Simethis planifolia
Sinapis arvensis, charlock
Sisyrinchium bermudiana, blue-eyed grass
Sorbus aucuparia, rowan
Spergula arvensis, corn spurrey
Spergularia rupicole, rock spurrey
Sphagnum, moss
Spiranthes romanzoffiana, threefold lady' s tresses
Stachys t., woundwort
Succisa pratensis, devil' s bit
Tilia, lime
Triticum dicoccum, emmer
Triticum monococcum, einkorn
Ulex gallii, autumn gorse
Ulmus, elm
Urtica, nettle
Vaccinium myrtillus, bilberry
Viscum, mistletoe
Zea mays, maize

BIBLIOGRAPHY

Abbreviations

BAR	British Archaeological Reports
JAS	Journal of Archaeological Science
JCHAS	Journal of the Cork Historical and Archaeological Society
JKAHS	Journal of the Kerry Archaeological and Historical Society
JRSAI	Journal of the Royal Society of Antiquaries of Ireland
PPS	Proceedings of the Prehistoric Society
PRIA	Proceedings of the Royal Irish Academy
UJA	Ulster Journal of Archaeology

Ammerman, A. J. and L. L. Cavalli-Sforza 1973. 'A population model for the diffusion of early farming in Europe'. In C. Renfrew (ed.), The Explanation of Culture Change: Models in Prehistory, 343-357. Duckworth.

Andersen, S. Th. 1973. 'The differential pollen productivity of trees and its significance for the interpretation of a pollen diagram from a forested region'. In H. J. B. Birks and R. G. West (eds.), Quaternary Plant Ecology, 109-115. Blackwell, Oxford.

Andersen, S. Th. 1979. 'Brown earth and podzol: soil genesis illuminated by microfossil analysis', BOREAS 8, 59-73.

ApSimon, A. 1976. 'Ballynagilly and the beginning and end of the Irish Neolithic'. In S. J. De Laet (ed.), Acculturation and Continuity in Atlantic Europe mainly during the Neolithic period and the Bronze Age, 15-30. de Tempel, Brugge.

Barber, J. 1972. The Stone Circles of Cork and Kerry: A study. Unpublished M. A. thesis, University College, Cork.

Barber, J. 1973. 'The orientation of the recumbent-stone circles of the south-west of Ireland', JKAHS, 6, 26-39.

Barber, J. in preparation (a) Excavations at Cashelkeelty, Co. Kerry.

Barber, J. in preparation (b) Excavation at Tor Mor, Isle of Arran.

Beckers, A., Th. Brock and J. Klerkx 1976. A Vegetation of some Parts of Dooaghtry, Co. Mayo, Republic of Ireland. Published privately by the Laboratory for Geobotany, Catholic University, Nijmegen, the Netherlands.

Beug, H. J. 1961. Leitfaden der Pollenbestimmung für Mitteleuropa und angrenzende Gebiete. Gustav Fischer, Stuttgart.

Birks, H. H. 1972. 'Studies in the vegetational history of Scotland II. Two pollen diagrams from the Galloway Hills, Kirkcudbrightshire', Journal of Ecology 60, 182-217.

Blytt, A. 1876. Essay on the immigration of the Norwegian flora during alternating rainy and dry periods. Christiana.

Boserup, E. 1965. The Conditions of Agricultural Growth. Allen and Unwin, London.

Boulet, L. J. 1939. The Ecology of a Welsh Mountain Sheep-Walk. Unpublished Ph. D. thesis, University of Wales.

Bowen, E. G. 1970. 'Britain and the British Seas'. In D. Moore (ed.), The Irish Sea Province in Archaeology and History, 13-28. Cardiff.

Bradley, R. 1978. The Prehistoric Settlement of Britain. Routledge and Kegan Paul.

Bradley, R. and I. Hodder 1979. 'British prehistory: an integrated view', Man 14, 1, 93-104.

Burenhult, G. 1980. The Archaeological Excavation at Carrowmore, Co. Sligo, Ireland Excavation Seasons 1977-79. Theses and Papers in North-European Archaeology, No. 9. University of Stockholm.

Butzer, R. W. 1964. Environment and Archaeology, An Introduction to Pleistocene Geography. Aldine, Chicago.

Case, H. 1966. 'Were Beaker-people the first metallurgists in Ireland?', Palaeohistoria 12, 141-77.

Case, H. 1969a. 'Neolithic Explanations', Antiquity 43, 176-86.

Case, H. 1969b. 'Settlement-patterns in the North Irish Neolithic', UJA 32, 3-27.

Caulfield, S. 1978. 'Neolithic Fields — the Irish Evidence'. In H. C. Bowen and P. J. Fowler (eds.) Early Land Allotment. BAR 48. Oxford.

Clark, R. M. 1975. 'A calibration curve for radiocarbon dates', Antiquity 69, 251-266.

Clarke, D. L. 1978. Analytical Archaeology. Methuen, London.

Coles, J. M. 1967. 'Some Irish horns of the late Bronze Age', JRSAI 97, 113-7.

Conry, M. J. 1970. Man's Role in Soil Profile Modification and Formation in Ireland. Unpublished Ph. D. thesis, Trinity College, Dublin.

Conry, M. and P. Ryan 1963. 'Soils'. In West Cork Resource Survey. An Foras Talúntais, Dublin.

Cornwall, I. W. 1953. 'Soil science and archaeology with illustrations from some British Bronze Age monuments', PPS 19, 129-147.

Damon, P. E., A. Long and E. I. Wallick 1972. 'Dendrochronologic calibration of the carbon-14 timescale'. In T. A. Rafter and T. Grant-Taylor (eds.) Proc. 8th Int. Conference on Radiocarbon Dating, 1, A27-43.

Davies, O. 1938. 'Castledamph stone circle', JRSAI 68, 106-112.

Dennell, R. W. 1976. 'Prehistoric crop cultivation in Southern England: a reconsideration', The Antiquaries Journal 56, 1, 11-23.

de Jong, A. F. M. and W. G. Mook 1979. 'Confirmation of the Suess wiggles: 3200-3700 BC', Nature 280, No. 5717, 48-49.

de Paor, M. and L. 1978. Early Christian Ireland. Thames and Hudson, London.

de Valera, R. 1979. In S. P. O Riordain, Antiquities of the Irish Countryside, 1-28. 5th edition. Methuen, London.

de Valera, R. and S. Ó Nualláin 1961. Survey of the Megalithic Tombs of Ireland. 1, County Clare. Stationery Office, Dublin.

de Valera, R. and S. O Nualláin 1964. Survey of the Megalithic Tombs of Ireland, 2, County Mayo. Stationery Office, Dublin.

de Valera, R. and S. O Nualláin 1972. Survey of the Megalithic Tombs of Ireland, 3, Counties Galway, Roscommon, Leitrim, Longford, Westmeath, Laoighis, Offaly, Kildare, Cavan. Stationery Office, Dublin.

Dimbleby, G. W. 1957. 'Pollen analysis of terrestrial soils', New Phytologist, 56, 12-28.

Dimbleby, G. W. 1961. 'Soil pollen analysis', J. Soil Science 12, 1-11.

Dimbleby, G. W. 1962. Development of British Heathlands and their Soils. Oxford Forestry Memoirs No. 23.

Dimbleby, G. W. 1965. 'Post-Glacial changes in soil profiles', Proc. Royal Society B, 161, 984, 355-362.

Dresser, P. Q. 1970. A Study of Sampling and Pretreatments of Materials for Radiocarbon Dating. Unpublished Ph. D. thesis, Queen's University, Belfast.

Edwards, K. J. 1979. 'Palynological and temporal inference in the context of prehistory, with special reference to the evidence from lake and peat deposits', JAS 6, 255-270.

Eogan, G. 1964. 'The Later Bronze Age in Ireland in the light of recent research', PPS 30, 268-351.

Erdtman, G. 1924. 'Mitteilungen über einige irische Moore', Svensk Botanisk Tidskrift 18, 451-59.

Erdtman, G. 1927. 'Traces of history of forests of Ireland', The Irish Naturalists Journal, 1, 1-4.

Erdtman, G. 1928. 'Studies in the postarctic history of the forests of North-western Europe. I. Investigations in the British Isles'. Geologiska Föreningens i Stockholm Förhandlingar 50, 123-92.

Erdtman, G. B. Berglund and J. Praglowski 1961. 'An introduction to a Scandinavian pollen flora', Grana Palynologica 2:3, 3-92.

Evans, E. 1957. Irish Folk Ways. Routledge and Kegan Paul.

Faegri, K. and J. Iversen 1977. Textbook of Pollen Analysis. Munksgaard, Copenhagen.

Fahy, E. M., 1959. 'A recumbent-stone circle at Drombeg, Co. Cork', JCHAS 64, 1-27.

Fahy, E. M. 1960. 'A hut and cooking places at Drombeg, Co. Cork', JCHAS 65, 1-17.

Fahy, E. M., 1961. 'A Stone Circle, hut and dolmen at Bohonagh, Co. Cork', JCHAS 66, 93-104.

Fahy, E. M., 1962. 'A recumbent-stone circle at Reenascreena South, Co. Cork', JCHAS 67, 59-69.

Fahy, E. M. 1969. 'Early settlement in the Skibbereen area', JCHAS 74, 147-156.

Fleming, A. 1975. 'The implications of calibration'. In T. Watkins (ed.), Radiocarbon: Calibration and Prehistory, 101-8. Edinburgh University Press.

Freeman, T. W. 1972. Ireland. Methuen, London.

Gardiner, M. J. and P. Ryan. 1969. 'A new generalised soil map of Ireland and its land use interpretation', Irish Journal of Agric. Research 8, 95-109.

Gogan, L. S. 1931. 'A small stone circle at Muishire Beag', JCHAS 36, 9-19.

Goddard, A. 1971. Studies of the vegetational changes associated with initiation of blanket peat accumulation in NE Ireland. Unpublished Ph.D. thesis, Queen's University, Belfast.

Godwin, H. 1968. 'Studies of the Post-Glacial history of British vegetation: XV, organic deposits of Old Buckenham Mere, Norfolk', New Phytologist 67, 95-107.

Godwin, H. 1975. History of the British Flora. A Factual Basis for Phytogeography. Cambridge University Press.

Gordon, A. D. and H. J. B. Birks 1972. 'Numerical methods in Quaternary palaeoecology: I, zonation of pollen diagrams', New Phytologist 71, 961-79.

Gordon, A. D. and H. J. B. Birks 1974. 'Numerical methods in Quaternary palaeoecology: II comparison of pollen diagrams', New Phytologist 73, 221-249.

Greene, D. 1945. 'Early Irish society'. In M. Dillon (ed.), Early Irish Society. Dublin.

Groenman-van Waateringe, W. 1968. 'The elm decline and the first appearance of Plantago major', Vegetatio 15, 292-296.

Grohne, U. 1957. 'Die Bedeutung des Phasenkontrastverfahrens für die Pollenanalyse, dargelegt am Beispiel der Gramineenpollen vom Getreidetyp', Photographie und Forschung 7, 8, 237-248.

Harbison, P. 1973. 'The Earlier Bronze Age in Ireland', JRSAI 103, 93-152.

Harbison, P. 1979. 'Who were Ireland's first Metallurgists?'. In M. Ryan (ed.) The Origins of Metallurgy in Atlantic Europe. Proc. 5th Atlantic Colloquium, Dublin.

Havinga, A. J. 1971. 'An experimental investigation into the decay of pollen and spores in various soil types'. In J. Brooks et. al., (eds.), Sporopollenin, 446-79. Academic Press.

Havinga, A. J. 1974. 'Problems in the interpretation of pollen diagrams of mineral soils', Geologie en Mijnbouw 53 (6), 449-453.

Herity, M. 1970. 'The prehistoric peoples of Kerry: a programme of investigation', JKAHS 3, 4-14.

Herity, M. and G. Eogan 1977. Ireland in Prehistory. Routledge and Kegan Paul.

Heybroek, H. M. 1963. 'Diseases and lopping for fodder as possible causes of a prehistoric decline of Ulmus', Acta Bot. Neerl. 12, 1-11.

Higgs, E. S. and C. Vita-Finzi 1972. 'Prehistoric economies: a territorial approach'. In E. S. Higgs (ed.), Papers in Economic Prehistory. Cambridge University Press.

Hughes, K. and A. Hamlin 1977. The Modern Traveller to the Early Irish Church. SPCK, London.

Iversen, J. 1944. 'Viscum, Hedera and Ilex as climatic indicators', Geologisk Forenings Forhandlinger 66, 3.

Iversen, J. 1960. 'Problems of early Post-Glacial forest development in Denmark', Geological Survey of Denmark 4, 1-32.

Jackson, J. 1979. 'Metallic ores in Irish prehistory: copper and tin'. In M. Ryan (ed.), The Origins of metallurgy in Atlantic Europe. Proc. 5th Atlantic Colloquium, Dublin.

Janssen, C. R. 1967. 'A comparison between the recent regional pollen rain and the subrecent vegetation in four major vegetation types in Minnesota (U.S.A.)', Review of Palaeobotany and Palynology 2, 331-342.

Janssen, C. R. 1970. 'Problems in the recognition of plant communities in pollen diagrams', Vegetatio 20, 187-198.

Jessen, K. 1949. 'Studies in Late Quaternary deposits and flora-history of Ireland', PRIA 52B, 85-290.

Jessen, K. and H. Helbaek 1944. 'Cereals in Great Britain and Ireland in prehistoric and early historic times', Det Kongelige Danske Videnskabernes Selskab, Biologiske Skrifter, III, 2. København.

Jonassen, H. 1950. 'Recent pollen sedimentation and Jutland heath diagrams', Dansk Bot. Ark. 13, 1-168.

Kavanagh, R. M. 1973. 'The encrusted urn of Ireland', PRIA 73, 507-617.

Kirk, S. M. 1973-74. 'High altitude cereal growing in County Down, Northern Ireland? A note', UJA 36-37, 99-100.

Lamb, H. H. 1974. 'Climate, vegetation and forest limits in early civilized times'. In F. R. Hodson (ed.), The Place of Astronomy in the Ancient World, 195-230. Oxford University Press.

L'Helgouach, J. 1973. 'Les mégalithes de l'ouest de la France. Evolution et chronologie'. In G. Daniel and P. Kjaerum (eds.). Megalithic Graves and Ritual, 203-220. Proc. of 3rd Atlantic Colloquium, Moesgard.

Limbrey, S. 1975. Soil Science and Archaeology. Academic Press.

Ní Loingsigh, A. (A. Lynch) 1976. The stone alignments of Cork and Kerry. Unpublished M. A. thesis, University College, Cork.

Lynch, A. in press. 'Excavation of a stone alignment at Maughanasilly, Bantry, Co. Cork', Cingula 1981.

Martin, F. X. 1967. 'The Anglo-Norman Invasion'. In T. W. Moody and F. X. Martin (eds.), The Course of Irish History, 123-143, Mercier Press, Cork.

Mc Aulay, R. and W. A. Watts 1961. 'Dublin radiocarbon dates I', Radiocarbon 3, 26-38.

Mc Cracken, E. 1971. The Irish Woods since Tudor Times. David and Charles.

Mc Kerrell, H. 1975. 'Correction procedures for C-14 dates'. In T. Watkins (ed.), Radiocarbon: Calibration and Prehistory 47-100. Edinburgh University Press.

Mitchell, G. F. 1945. 'The relative ages of archaeological objects recently found in bogs in Ireland', PRIA 50C, 1, 1-19.

Mitchell, G. F. 1951. 'Studies in Irish Quaternary deposits', PRIA 53B, 11-206.

Mitchell, G. G. 1956. 'Post-boreal pollen diagrams from Irish raised bogs', PRIA 57B, 14, 185-251.

Mitchell, G. F. 1965. 'Littleton Bog, Tipperary: an Irish agricultural record', Geol. Soc. Amer. Special paper 84, 121-132.

Mitchell, G. F. 1972. 'Soil deterioration associated with prehistoric agriculture in Ireland', Proc. 24th Int. Geol. Congr., Montreal Symp., 1, 59-68.

Mitchell, G. F. 1976. The Irish Landscape. Collins, London.

Moore, P. D. 1973. 'The influence of prehistoric cultures upon the initiation and spread of blanket bog in upland Wales', Nature 241, 350-353.

Moore, P. D. 1974. 'Prehistoric human activity and blanket peat initiation on Exmoor', Nature 250, 439-441.

Moore, P. D. 1975. 'Origin of blanket mires', Nature 256, 267-269.

Moore, P. D. and J. A. Webb 1978. An Illustrated Guide to Pollen Analysis. Hodder & Stoughton.

Morrison, M. E. S. 1959. 'Evidence and interpretation of "landnam" in the North-East of Ireland', Botanisker Notiser 112, 185-204.

O Cuív, B. 1967. 'Ireland in the eleventh and twelfth centuries'. In T. W. Moody and F. X. Martin (eds.), The Course of Irish History. Mercier Press, Cork.

O Corráin, D., forthcoming New History of Ireland vol. I.

O Drisceoil, D. 1980. Fulachta Fiadh: A Study. Unpublished M. A. thesis, University College, Cork.

O'Kelly, M. J. 1951. 'An early bronze age ring-fort at Carrigillihy, Co. Cork', JCHAS 56, 69-86.

O'Kelly, M. J. 1952. 'Three promontory forts in Co. Cork', PRIA 55C, 26-59.

O'Kelly, M. J. 1954. 'Excavations and experiments in ancient Irish cooking places', JRSAI 84, 105-155.

O'Kelly, M. J. 1962. 'Two ring-forts at Garryduff, Co. Cork', PRIA 63C, 17-124.

O'Kelly, M. J. 1969. 'A stone mould for axeheads from Doonour, Bantry, Co. Cork', JRSAI 99 Part 2, 117-124.

O'Kelly, M. J. forthcoming 'The megalithic tombs of Ireland'. In forthcoming Glyn Daniel Festschrift.

O'Kelly, M. J. and E. Shee 1974. 'Bronze age burials at Coolnahane and Ballinvoher, Co. Cork', JCHAS 80, 71-85.

O Nualláin, S. 1975. 'The stone circle complex of Cork and Kerry', JRSAI 105, 83-131.

O Nualláin, S. 1979. 'The megalithic tombs of Ireland', Expedition 21, No. 3, 6-15.

O Riordáin, S. P. 1939. 'Excavation of a stone circle and cairn at Kealkil, Co. Cork', JCHAS 44, 46-49.

O Riordáin, S. P. 1942. 'The excavation of a large earthen ring-fort at Garranes, Co. Cork', PRIA 47C, 77-150.

O Riordáin, S. P. and P. Hartnett 1943. 'The excavation of Ballycatteen Fort, Co. Cork', PRIA 54C, 37-74.

O' Shea, J. 1980. 'Social Storage'. Lecture delivered at Cambridge conference, March 1980.

O' Sullivan, P. E. 1976. 'Pollen analysis and radiocarbon dating of a core from Lough Pityoulish, Eastern Highlands of Scotland', Journal of Biogeography 3, 293-302.

Oldfield, F. 1963. 'Pollen analysis and man's role in the ecological history of the south-east Lake District', Geogr. Annlr. 45, 1, 23-40.

Orme, A. R. 1970. The World's Landscapes — Ireland. Longman.

Pearson, G. W. 1979. 'Belfast radiocarbon dates IX', Radiocarbon 21, No. 2, 274-290.

Pearson, G. W. and J. R. Pilcher 1975. 'Belfast radiocarbon dates VIII', Radiocarbon 17, No. 2, 226-238.

Pearson, G. W., J. R. Pilcher, M.G. L., Baillie and J. Hillam 1977. 'Absolute radiocarbon dating using a low altitude European tree-ring calibration', Nature 270, 25-28.

Pennington, W. 1970. 'Vegetation history in the north-west of England: a regional synthesis'. In D. Walker and R. G. West (eds.), Studies in the Vegetational History of the British Isles, 41-79. Cambridge University Press.

Pennington, W. 1975. 'The effect of Neolithic man on the environment in northwest England: the use of absolute pollen diagrams'. In J. G. Evans, S. Limbrey and H. Cleere (eds.), The Effect of Man on the Landscape. The Highland Zone 74-86. C. B. A. Research Report No. 11.

Pennington, W., E. Y. Haworth, A. P. Bonny and J. P. Lishnam 1972. 'Lake sediments in northern Scotland', Philosophical Transactions of the Royal Society B264, 191-294.

Pilcher, J. R. 1968. 'Some applications of scanning electron microscopy to the study of modern and fossil pollen', UJA 31, 87-91.

Pilcher, J. R. 1969. 'Archaeology, palaeoecology and C^{14} dating of the Beaghmore stone circle site', UJA 32, 73-91.

Pilcher, J. R.. 1970. Palaeoecology and the Radiocarbon Dating of Sites in Co. Tyrone, Northern Ireland. Unpublished Ph.D thesis, Queen's University, Belfast.

Pilcher, J. R. 1973. 'Pollen analysis and radiocarbon dating of a peat at Slieve Gallion, Co. Tyrone, N. Ireland', New Phytol. 72, 681-689.

Pilcher, J. R., A. G. Smith, G. W. Pearson and A. Crowder 1971. 'Land clearance in the Irish Neolithic: new evidence and interpretation', Science 172, 560-562.

Pilcher, J. R. and A. G. Smith 1979. 'Palaeoecological investigations at Ballynagilly, a Neolithic and Bronze Age settlement in County Tyrone, Northern Ireland', Philosophical Transactions of the Royal Society 286, 345-369.

Praeger, R. LL. 1932a. 'Some noteworthy plants found in or reported from Ireland', PRIA 41B, 95-124.

Praeger, R. LL. 1932b. 'Recent views bearing on the problem of the Irish flora and fauna', PRIA 41B, 125-145.

Praeger, R. LL. 1977. The Botanist in Ireland. EP, Wakefield (originally published 1934).

Proudfoot, V. B. 1958. 'Problems of soil history. Podzol development at Goodland and Torr townlands, Co. Antrim, Northern Ireland', Journal of Soil Science 9, 186-198.

Proudfoot, V. B. 1961. 'The economy of the Irish rath', Medieval Archaeology 5, 94-121.

Proudfoot, B. 1970. 'Irish raths and cashels: some notes on chronology, origins and survivals', UJA 33, 37-48.

Raftery, B. 1972. 'Irish hill-forts'. In C. Thomas (ed.), The Iron Age in the Irish Sea Province, 37-58. C.B.A. Research Report No. 9.

Raftery, B. 1976. 'Rathgall and Irish hillfort problems'. In D. W. Harding (ed.), Hillforts, Later Prehistoric Earthworks in Britain and Ireland, 339-358. Academic Press.

Ralph, E. K., H. N. Michael and M. C. Han 1973. 'Radiocarbon dates and reality', MASCA Newsletter 9, No. 1, 1-20.

Rees, S. 1979. Agricultural Implements in Prehistoric and Roman Britain. BAR 69. Oxford.

Roberts, B. K. J. Turner and P. F. Ward 1973. 'Recent forest history and land use in Weardale, Northern England'. In H. J. B. Birks and R. G. West (eds.), Quaternary Plant Ecology, 207-221. Blackwell, Oxford.

Rohan, P. K. 1975. The Climate of Ireland. Stationery Office, Dublin.

Romans, J. C. C. 1970. 'Podzolisation in a zonal and altitudinal context in Scotland', Welsh Soils Discussion Group No. 11, 88-101.

Scannell, M. J. P. and D. M. Synnott 1972. Census Catalogue of the Flora of Ireland. Stationery Office, Dublin.

Scott, B. G. 1979. 'The introductions of non-ferrous and ferrous metal technologies to Ireland: motives and mechanisms'. In M. Ryan (ed.), The Origins of Metallurgy in Atlantic Europe, 189-202. Proc. of 5th Atlantic Colloquium, Dublin.

Sernander, R. 1908. 'On the evidence of postglacial changes of climate, furnished by the peat-mosses of northern Europe', Geologiska Foreningens Förhandlinger 30, 465-478.

Simmons, I. G. 1964. 'Pollen diagrams from Dartmoor', New Phytol. 63, 165-180.

Simmons, I. G. 1969. 'Evidence for vegetation changes associated with Mesolithic man in Britain'. In P. J. Ucko and G. W. Dimbleby (eds.), The Domestication and Exploitation of Plants and Animals, 111-119. Duckworth.

Simmons, I. G. and Cundhill, P. R. 1974. 'Late Quaternary vegetational history of the North York Moors', Journal of Biogeography 1, 159-169.

Sims, R. E. 1973. 'The anthropogenic factor in East Anglian vegetational history: an approach using A. P. F. techniques'. In H. J. B. Birks and R. G. West (eds.) Quaternary Plant Ecology, 81-96. Blackwell, Oxford.

Sims, R. E. 1978. 'Man and vegetation in Norfolk'. In S. Limbrey and J. G. Evans (eds.). The Effect of Man on the Landscape: the Lowland Zone, 57-63. C.B.A. Research Report No. 21.

Smith, A. G. 1961a. 'The atlantic-sub-boreal transition', Proc. of the Linnean Society of London 172, 38-49.

Smith, A. G. 1961b. 'Cannons Lough, Kilrea, Co. Derry: stratigraphy and pollen analysis', PRIA 61B, 369-383.

Smith, A. G. 1964. 'Problems in the study of the earliest agriculture in Northern Ireland', Rep. VIth Int. Cong. Quat. 2, 461-471. Warsaw.

Smith, A. G. 1970a. 'Late- and post-glacial vegetational and climatic history of Ireland: a review', Irish Geographical Studies, 65-88.

Smith, A. G. 1970b. 'The influence of Mesolithic and Neolithic man on British vegetation: a discussion'. In D. Walker and R. G. West (eds.), Studies in the Vegetational History of the British Isles, 81-96. Cambridge University Press.

Smith, A. G. 1975. 'Neolithic and Bronze age landscape changes in northern Ireland'. In J. G. Evans, S. Limbrey and H. Cleere (eds.), The Effect of Man on the Landscape: the Highland Zone, 64-74. C.B.A. Research Report No. 11.

Smith, A. G. and E. H. Willis 1961-62. 'Radiocarbon dating of the Fallahogy landnam phase', UJA 24-25, 16-24.

Smith, A. G., G. W. Pearson and J. R. Pilcher 1970. 'Belfast radiocarbon dates I', Radiocarbon 12, No. 1, 285-290.

Smith, A. G., G. W. Pearson and J. R. Pilcher 1971. 'Belfast radiocarbon dates III', Radiocarbon 13, No. 1, 103-125.

Smith, A. G. and J. R. Pilcher 1973. 'Radiocarbon dates and vegetational history of the British Isles', New Phytol. 72, 903-914.

Spratt, D. A. and I. G. Simmons 1976. 'Prehistoric activity and environment on the North York Moors', JAS 3, 193-210.

Suess, H. E. 1965. 'Secular variations in the cosmic-ray-produced carbon-14 in the atmosphere', J. Geophysical Res. 70, 5937-52.

Suess, H. E. 1967. 'Radioactive dating and methods of low level counting', IAEA, 143-151. Vienna.

Suess, H. E. 1978. 'La Jolla measurements of radiocarbon in tree-ring dated wood', Radiocarbon 20, 1-18.

Tallis, J. H. 1964. 'Studies on southern Pennine peats III. The behaviour of Sphagnum', Journal of Ecology 52, 345-353.

Tansley, A. G. 1939. The British Isles and their Vegetation. Cambridge.

Ten Hove, H. A. 1968. 'The Ulmus fall at the transition atlanticum — subboreal in pollen diagrams', Palaeogeography, Palaeoclimatology, Palaeoecology 5, 359-369.

Teunissen, D. and H. G. C. M. Teunissen-van Oorschot 1980. 'The history of the vegetation in SW Connemara (Ireland)', Acta Bot. Neerl. 29 (4), 285-306.

Tinsley, H. M. and R. T. Smith 1974. 'Surface pollen studies across a woodland/heath transition and their application to the interpretation of pollen diagrams', New Phytol. 73, 547-565.

Troels-Smith, J. 1954. 'Ertebøllekultur- Bondekultur (Ertebølle Culture — Farmer Culture)', Aarb Nord. Oldk. Hist. 1953, 1-62. Copenhagen.

Troels-Smith, J. 1960. 'Ivy, mistletoe and elm. Climatic indicators — fodder plants. A contribution to the interpretation of the pollen zone border VII-VIII', Geological Survey of Denmark 4, 1-31.

Turner, J. 1964. 'The anthropogenic factor in vegetational history: I. Tregaron and Whixall Mosses', New Phytol. 63, 73-90.

Van Wijngaarden-Bakker, L. H. 1974. 'The animal remains from the beaker settlement at Newgrange, Co. Meath: first report', PRIA 74C, 313-383.

Van Wijngaarden-Bakker, L. H. forthcoming. 'The animal remains from the beaker settlement at Newgrange, Co. Meath: final report', PRIA 1981.

Voorrips, A. 1973. 'An ALGOL-60 program for computation and graphical representation of pollen analytical data', Acta Bot. Neerl. 22 (6), 645-654.

Voorrips, A. 1974. 'An ALGOL-60 program for pollen analytical data: the CDC version', Acta Bot. Neerl. 23 (5)-(6), 701-704.

Voorrips, A. in preparation. 'A simple method for the comparison of pollen diagrams'.

Waddell, J. 1970. 'Irish Bronze age cists: a survey', JRSAI 100, 91-139.

Wailes, B. 1972. 'Plow and population in temperate Europe'. In B. Spooner (ed.), Population Growth: Anthropological Implications, 154-179. MIT Press.

Wainwright, G. J. 1972. 'The excavation of a Neolithic settlement on Broome Heath, Ditchingham, Norfolk', PPS 38, 1-97.

Walker, D. 1966. 'The late Quaternary history of the Cumberland Lowland', Philosophical Transactions of the Royal Society B251, 1-210.

Walker, D. 1970. 'Direction and rate in some British post-glacial hydroseres'. In D. Walker and R. G. West (eds.), Studies in the Vegetational History of the British Isles, 117-140. Cambridge University Press.

Wallace, P. F. 1977. 'A prehistoric burial cairn at Ardcrony, Nenagh, Co. Tipperary', North Munster Antiquarian Journal 19, 3-20.

Waterbolk, H. T. 1957. 'Pollenanalytisch onderzoek van twee noordbrabantse tumuli', (in G. Beex, 'Twee grafheuvels in Noord-Brabant), Bijdr. Studie Brabantse Heem. 9, 34-39.

Waterbolk, H. T. 1968. 'Food production in prehistoric Europe', Science 162, 1093-1102.

Waterman, D. 1964. 'The stone circle, cairn and alignment at Drumskinny, Co. Fermanagh' UJA 27, 23-30.

Webb, D. A. 1977. An Irish Flora. Dundalgan Press.

Welten, M. 1952. 'Pollenanalytische Stichproben über die subrezente Vegetationsentwicklung im Bergland von Kerry (Irland)'. Die Pflanzenwelt Irlands (The Flora and Vegetation of Ireland). Ergebnisse der 9. Internationalen Pflanzengeographischen Exkursion durch Irland 1949. Bern.

West, R. G. 1970 'Pollen zones in the pleistocene of Great Britain and their correlation', New Phytol. 69, 1179-1183.

Whittle, A. W. R. 1977. The Earlier Neolithic of S. England and its Continental Background. B A R Supplementary Series 35. Oxford.

Woodman, P. C. 1974. 'Settlement patterns of the Irish Mesolithic', UJA 36/37, 1-16.

Woodman, P. C. 1976. 'The Irish Mesolithic/Neolithic transition'. In S. J. De Laet (ed.), Acculturation and continuity in Atlantic Europe mainly during the Neolithic Period and the Bronze Age, 296-307. de Tempel, Brugge.

Woodman, P. C. 1978. The Mesolithic in Ireland: Hunter Gatherers in an Insular Environment. BAR 58. Oxford.

www.ingramcontent.com/pod-product-compliance
Lightning Source LLC
LaVergne TN
LVHW072256300926
849160LV00018B/1676